Advance Praise for

Earthbag Building
The Tools, Tricks and Techniques

This inviting, complete guide to earthbag construction is humorous, very well written, and chock
full of good ideas and dynamite illustrations. When you finish reading this book
there's only one thing left to do: get out there and get to it!

— Dan Chiras, Co-author of *The Natural Plaster Book* and author of *The Natural House,*
The Solar House, and *Superbia! 31 Ways to Create Sustainable Neighborhoods*

Natural building practitioners, like Kaki and Doni, have persevered through years of trial and error,
teaching, learning, innovating and becoming respected leaders of the natural
building community. As *Earthbag Building: The Tools, Tricks and Techniques* demonstrates,
Kaki and Doni are smart, they are playful, they are wise, they are fine teachers and they
have lots of get down and dirty practical experience to share about how to transform bags of earth
and earth/lime plasters into beautiful and sensual buildings. We offer a deep bow
to these champions of natural building, who (we now know) are doing *real* and
transformational work; offering us doable ways to meet our basic human need for
shelter in ways that are restorative and sustainable to both the earth and the spirit.

— Judy Knox and Matts Myhrman, Out On Bale, Tucson, Arizona

Who would have thought that you could make a beautiful, super solid and durable home using
dirt-filled grain sacks? *Earthbag Building* shows not only that you can,
but that you can have fun and feel secure doing it. With humor, integrity and delight,
Kaki and Doni have distilled into written word and clear illustration their years of
dedicated research and work refining the process and tools for this promising
building technique. Their thorough approach and objective discussions
of pros, cons and appropriate applications makes this book a must-read for
natural building enthusiasts and skeptics alike.

— Carol Escott and Steve Kemble, co-producers of
How To Build Your Elegant Home with Straw Bales

Earthbag

BUILDING

The Tools, Tricks and Techniques

Kaki Hunter and
Donald Kiffmeyer

NEW SOCIETY PUBLISHERS

Cataloguing in Publication Data:
A catalog record for this publication is available from the National Library of Canada.

Cover design by Diane McIntosh. Cover Image: Kaki Hunter and Donald Kiffmeyer.

Printed in Canada. Tenth printing, January 2022.

Paperback ISBN: 978-0-86571-507-3

Inquiries regarding requests to reprint all or part of *Earthbag Building* should be addressed to New Society Publishers at the address below.

To order directly from the publishers, please add $4.50 shipping to the price of the first copy, and $1.00 for each additional copy (plus GST in Canada). Send check or money order to:

New Society Publishers
P.O. Box 189, Gabriola Island, BC V0R 1X0, Canada
1-800-567-6772

New Society Publishers' mission is to publish books that contribute in fundamental ways to building an eco-logically sustainable and just society, and to do so with the least possible impact on the environment, in a manner that models this vision. We are committed to doing this not just through education, but through action. We are acting on our commitment to the world's remaining ancient forests by phasing out our paper supply from ancient forests worldwide. This book is one step towards ending global deforestation and climate change. It is printed on acid-free paper that is **100% old growth forest-free** (100% post-consumer recycled), processed chlorine free, and printed with vegetable based, low VOC inks. For further information, or to browse our full list of books and purchase securely, visit our website at: **www.newsociety.com**

NEW SOCIETY PUBLISHERS   www.newsociety.com

Books for Wiser Living from Mother Earth News

Today, more than ever before, our society is seeking ways to live more conscientiously. To help bring you the very best inspiration and information about greener, more-sustainable lifestyles, New Society Publishers has joined forces with *Mother Earth News*. For more than 30 years, *Mother Earth* has been North America's "Original Guide to Living Wisely," creating books and magazines for people with a passion for self-reliance and a desire to live in harmony with nature. Across the countryside and in our cities, New Society Publishers and *Mother Earth News* are leading the way to a wiser, more sustainable world.

Contents

Acknowledgments

Right off the bat, we'd like to thank Chris Plant at NSP for his perseverance, patience and persistence in pursuing his interest in our book project ever since that fateful phone call in 2000. Yep folks, that's how long ago we started this mission. Constructing *Earthbag Building* has been a monumental undertaking, more so than actually building an earthbag house! But we now know that all the fret, sweat and zillion hours has turned a bunch of paper and ink into a dirtbag manifesto of beauty and usefulness ready to inspire alternative builders around the world. We are proud of our collective achievement. Thank you Chris for taking this on!

Kudos go to our editor, Ingrid Witvoet and Artistic Designer, Greg Green for plowing through the voluminous material we bombarded them with. Special thanks goes to Sue Custance for her steadfast participation and careful arrangement of the layout. It is no mean feat to fit some 480 plus images within 280 some pages.

Much appreciation goes to our local support system, Tom and Lori O'Keefe at Action Shots, Teresa King and company at Canyonlands Copy center and Dan Norris at Ancient Images.

With much love and gratitude we'd like to thank our families, Tom and Katherine Hunter (Kaki's parents) and Doni's mom Helen Kiffmeyer and his siblings Joe Kiffmeyer and Carolyn Schwier for their unwavering encouragement and our loyal friends for still loving us in spite of the many times we'd declined invitations to do fun stuff because, "...oh, man, we'd love to but ... we're still working on the book...(four years later) ... uh ... still working on the book ... the book ... still working on it ... yep, the same book..."

Thank you Boody Springer (Kaki's son) — you and your generation were a tremendous motivation for this work. Thank you Christy Williams, Elenore Hedden and Cynthia Aldrige for working your white magic on healing you know what in the nick of you know when.

A big fat hug goes to our partner in grime, (the third ok in okokok Productions), Kay Howe. She, more than anyone was (and still is) the most positive, personable, playful, proactive dirtbag enthusiast we know. While we were building the Honey House an onlooker commented, "That sure looks like a lot of hard work." Kay responded laughing, "So what?" (This attitude from a single mother of four).

Lastly, we'd like to thank everyone that has ever handed us a can of dirt, diddled a corner with us, tamped a row, hardassed a butt, played ring around the barbed wire or just plain stood around and made brilliant suggestions that we were too oblivious to notice, we'd like to say from the bottom of our hearts — Hurray! Thank God it's finished!!

We love you all sooooo much!
— Kaki Hunter and Doni Kiffmeyer

Foreword

BY LYNNE ELIZABETH

Building with earthbags is gutsy. Gutsy because only the brave take up a construction method so different from the conventional. Gutsy because people build homes with this technique when they've just learned it. Gutsy because the materials are basic, elemental, primal. And gutsy, indeed, because this construction system resembles, in form and assembly, nothing other than our own intestines!

A shovel, bags, a little barbed wire and the earth beneath are all that are needed to build with earthbags. The method offers more structural integrity than adobe, more plasticity than rammed earth, and more speed in construction than cob. Although earthbag is new compared to these ancient building methods, it offers superior economy and durability in domed and vaulted assemblies. Earthbag construction offers broad possibility for ultra-low-cost, low-impact housing, especially in regions where timber, grasses, cement, and fuel are scarce. Earthbag domes also provide unparalleled safety in wooded areas prone to wildfires, as fire will more easily pass over any structures without a roof or eaves to ignite. Earthbag building has been chosen, too, for sites exposed to hurricanes and other extreme weather. Solid as the earth itself, it holds great thermal mass and cannot rot or be eaten by insects.

Military bunkers and trenches were constructed with earthbags during World War I, and the use of sand or earthbag retaining walls to divert flood waters is ubiquitous. Appropriate building technologists Otto Frei and Gernot Minke of Germany experimented independently in the 1960s and 70s with wall systems using earth-filled bags.

Credit for developing contemporary earthbag construction goes to architects Nader Khalili and Illiona Outram of the California Earth Art and Architecture Institute in Hesperia, known as Cal-Earth. Starting with domed and vaulted assemblies of individual earth-packed bags, they later discovered that the polypropylene bags they had been stuffing could be obtained in uncut, unstitched, continuous tubes. With minor adjustments to the filling and assembly process, these long casings provided an efficient method to construct unbroken wall sections. Cal-Earth named these continuous bag assemblies "Superadobe" and, although descriptive names such as "flexible-form rammed earth" (adopted by this book's authors) and "modular contained earth" have been used, the most simple name — earthbag — still holds favor. It is, after all, a basic system.

Although Cal-Earth holds a United States patent for Superadobe construction, they share the technology freely, knowing that few other building methods are as ecological or as affordable. Their students have taken the method throughout the United States and other countries for two decades now, and several teach and have authored their own books on earth building. Joseph Kennedy brought earthbags to ecovillages in South Africa, and Paulina Wojciechowska brought the style to England, West Africa, and Europe. Earthbag structures have also been built in Mexico, Haiti, Chile, Brazil, Mongolia, and recently even by nuns in Siberia. The method is easily learned. With little training other than a site visit to Cal-Earth, artist Shirley Tassencourt built an earthbag meditation dome at age 69. She subsequently involved her grandson, Dominic Howes, in building an earthbag home, and Dominic went on to pioneer different earthbag structural forms in new climates, including Wisconsin.

Simple though it is in concept, the practice of earthbag building has been significantly refined by Kaki Hunter and Doni Kiffmeyer. This couple has moved earthbag construction out of a developmental era into one in which building contractors can be trained and building standards adopted. The uniform bag courses, tamping tools, and tidy bag corners of their Honey House, constructed a decade ago, showed for the first time that earthbag construction was ready to move into the mainstream. Kaki and Doni's continued attention to detail has advanced assembly techniques, and their meticulous documentation of earthbag building methods makes this book an ideal instruction manual for earthbag builders as well as a reference guide for building officials.

Earthbag was originally developed for self-help housing, and, true to that purpose, the techniques presented in this book are explained through photographs, line-drawings, and words in an easily understandable way. It offers valuable service as a field manual in many countries, with or without translation, although it would be a shame not to translate the lively text. In addition to carefully sharing everything they know about this construction method, Kaki Hunter and Doni Kiffmeyer bring a candor and sense of humor that speak volumes about the natural building spirit.

—Lynne Elizabeth, Director, New Village Press
Editor, "Alternative Construction: Contemporary
Natural Building Methods"

Introduction to Earthbag Building

We were perplexed. The headline in our local newspaper read, "Creating Affordable Housing Biggest Problem This Decade." To us, this was a mysterious statement. Until the last century, affordable housing had been created with little or no problem in our area for over a thousand years. The Four Corners region of the Southwestern U.S. was more populous 800 to 1,000 years ago than it is today. Ancient builders provided housing using the materials on hand. Stone, sticks, clay, sand, fiber, and some timbers were all they used to build modest-sized, comfortable dwellings for all the inhabitants. With modern methods and materials, why is it so difficult to provide enough housing for less people today?

Unfortunately for all of us, the answer lies within the question. Current laws require the use of manufactured materials, extracted as natural resources miles away, processed in yet another location, and then transported great distances to us. Naturally, this drives the price of building a home beyond the reach of most people.

At the time we met we had yet to become acquainted with earthbag architecture. From our many walks in the desert we discovered a lot of common interests: acting, a love of nature, storytelling and food, parallel spiritual philosophies, rafting, Native American architecture, and the joy of building. We visited ancient Indian ruins, fantasizing about the way they lived. Inspired by the enduring beauty of their building techniques, we began to explore how we too could build simple structures with natural earth for ourselves. We considered various forms of earthen building: adobe block, rammed earth, coursed adobe, poured adobe, cob, sod, etc. It seemed peculiar that in such a dry climate there is not a single adobe brickyard in our area. Yet adobe structures built around the turn of the 1900's still stood within the city limits.

While we could see the value of using regionally available indigenous material, not everyone shares our view. We all have different tastes and styles of expression. So our challenge was to combine the naturally abundant materials all around us with manufactured materials that are created in excess, and would have appeal to a more conventional mindset.

A friend turned us on to a now out of print earthen architectural trade magazine called *The Adobe Journal*. That's when we discovered the work of Nader Khalili. Nader was building monolithic dome-shaped structures with arches out of grain bags and tubes filled with dirt; any kind of dirt, even dry sand. He called it Sandbag/Superadobe/Superblock and he was working with the local building department conducting extensive tests concerning the building's ability to withstand load and wind shear, and resist earthquakes. Since then he has acquired permits for building residential and commercial structures, including a nature and science museum in one of the highest earthquake zones in the United States.

We signed up for a one-day workshop. Nader personally taught us how to build an arch using bricks and dry sand, and then using sandbags. We were invited to spend the night in one of the prototype domes under construction. We were hooked. We came home and started building walls.

We tried flopping bags every which way, stomping on them, banging them with various tamping devices. We experimented with varying the moisture contents, making makeshift bag stands, and different kinds of bags, tubes, soils, and techniques. Our project attracted a lot of attention and we found ourselves helping others to build privacy walls, benches, planters, and even a small dome. But all the while our focus seemed to be directed toward technique. The process became our priority. How could we neaten up the bags, take the slack out of them, tighten their derrière, and simplify the job overall? It soon became our mission to "turn a bag of dirt into a precision wall-building system." Hence, the Flexible-Form Rammed Earth technique evolved.

The Flexible-Form Rammed Earth technique is our contribution to earthbag building. We practice a particular brand of earthbag building that prioritizes ease of construction coupled with structural integrity inspired by FQSS principles. What is FQSS? We made a list of what fosters a productive yet playful work environment. The process has to be *Fun*. What helps make the job fun is that it flows *Quickly*, as long as we keep it *Simple*, and the results are *Solid*. So we adopted the FQSS stamp of approval: Fun, Quick, Simple, and Solid. The Flexible-Form Rammed Earth

technique has and continues to be developed according to this FQSS criterion. When the work becomes in any way awkward or sloppy, FQSS deteriorates into fqss: frustrating, quarrelsome, slow, and stupid. This prompts us to re-evaluate our tactics, or blow the whole thing off and have lunch. Returning refreshed often restores FQSS approval spontaneously. By demonstrating guidelines that effectively enhance the quality of earthbag construction, we hope to encourage a standard that aids the mainstream acceptance of this unique contemporary form of earthen architecture.

Throughout this work we often use synonymous terms to describe the same thing. For example, we intermix the use of the words earth, soil, dirt, and fill. They are all used to describe the magical mix of naturally occurring sand and clay, sometimes with the addition of fiber, and almost always in conjunction with some amount of water. Our intent is to inform, educate, and inspire earthbag construction in playful layman terms using written text and step-by-step, how-to illustrations.

The focus of this book is on sharing our repertoire of tools, tricks, and techniques that we have learned through trial and error, from friends, workshop participants, curious onlookers, ancient Indian nature spirits, and smartass apprentices who have all helped us turn a bag of dirt into a precision wall-building system that alerts the novice and experienced builder alike to the creative potential within themselves and the very earth beneath their feet.

CHAPTER 1

The Merits of Earthbag Building

With a couple rolls of barbed wire, a bale of bags, and a shovel one can build a magnificent shelter with nothing more than the earth beneath their feet. This is the premise that inspired the imagination of international visionary architect Nader Khalili when he conceived the idea of Sandbag Architecture. In his quest to seek solutions to social dilemmas like affordable housing and environmental degradation, Nader drew on his skills as a contemporary architect while exercising the ingenuity of his native cultural heritage. Monolithic earthen architecture is common in his native home of Iran and throughout the Middle East, Africa, Asia, Europe, and the Mediterranean. Thousands of years ago, people discovered and utilized the principles of arch and dome construction. By applying this ancient structural technology, combined with a few modern day materials, Nader has cultivated a dynamic contemporary form of earthen architecture that we simply call Earthbag Building.

1.1:
Using earthbags, a whole house, from foundation to walls to the roof, can be built using one construction medium.

3

MARLENE WULF

1.2: Marlene Wulf's earthbag dome under construction, deep in the woods of Georgia.

Simplicity

Earthbag Building utilizes the ancient technique of *rammed earth* in conjunction with woven bags and tubes as a *flexible form.* The basic procedure is simple. The *bags* or *tubes* are filled on the wall using a suitable pre-moistened earth laid in a mason style *running bond.* After a row has been laid, it is thoroughly compacted with hand tampers. Two strands of 4-point barbed wire are laid in between every row, which act as a "velcro mortar" cinching the bags in place. This provides exceptional tensile strength while allowing the rows to be stepped in to create *corbelled domes* and other unusual shapes (Fig. 1.1).

Walls can be linear, free form, or a perfect circle guided by the use of an architectural compass. Arched windows and doorways are built around temporary *arch forms* until the *keystone bags* are tamped in place. The finished walls then cure to durable cement-like hardness.

Simple, low cost foundations consist of a *rubble trench system,* or beginning the bag-work below ground with a *cement-stabilized* rammed earth mix for the stem walls. Many other types of foundation systems can be adapted to the climatic location and function of the structure.

Cut Barbed Wire Not Trees

We have the ability to build curvaceous, sensual architecture inspired by nature's artistic freedom while providing profound structural integrity. Earthbag construction enables the design of monolithic architecture using natural earth as the primary structural element. By monolithic architecture we mean that an entire structure can be built from foundation and walls to roof using the same materials and methods throughout. Corbelled earthbag domes foster the ultimate experience in sculptural monolithic design, simplicity, beauty, and dirt-cheap thrills. Earthbag domes designed with arch openings can eliminate 95 percent of the lumber currently used to build the average stick frame house (Fig. 1.2).

Conventional wood roof systems still eat up a lot of trees. This may make sense to those of us who dwell in forested terrain, but for many people living in arid or temperate climates, designing corbelled earthbag domes offers a unique opportunity for providing substantial shelter using the earth's most abundant natural resource, the earth itself. Why cut and haul lumber from the Northwest to suburban Southern California, Tucson, or Florida when the most abundant, versatile, energy efficient, cost effective, termite, rot and fire proof construction material is available right beneath our feet? Even alternative wall systems designed to limit their use of wood can still swallow up as much as 50 percent of that lumber in the roof alone. Earth is currently and has been the most used building material for thousands of years worldwide, and we have yet to run out.

Advantages of Earthbag Over Other Earth Building Methods

Don't get us wrong. We *love* earthen construction in all its forms. Nothing compares with the beauty of an adobe structure or the solidity of a rammed earth wall. The sheer joy of mixing and plopping cob into a sculptural masterpiece is unequalled. But for the first-and-only-time owner/builder, there are some distinct advantages to earthbag construction. Let's look at the advantages the earthbag system gives the "do-it-yourselfer" compared to these other types of earth building.

Adobe is one of the oldest known forms of earthen building. It is probably one of the best examples of the durability and longevity of earthen construction (Fig 1.3).

Adobe buildings are still in use on every continent of this planet. It is particularly evident in the arid and semi-arid areas of the world, but is also found in some of the wettest places as well. In Costa Rica, C.A., where rain falls as much as 200 inches (500 cm) per year, adobe buildings with large overhangs exist comfortably.

Adobe is made using a clay-rich mixture with enough sand within the mix to provide compressive strength and reduce cracking. The mix is liquid enough to be poured into forms where it is left briefly until firm enough to be removed from the forms to dry in the sun. The weather must be dry for a long enough time to accomplish this. The adobes also must be turned frequently to aid their drying (Fig. 1.4).

SOUTH WEST SOLAR ADOBE (SWSA)

1.3: *A freshly laid adobe wall near Sonoita, Arizona.*

They cannot be used for wall building until they have completely cured. While this is probably the least expensive form of earthen building, it takes much more time and effort until the adobes can be effectively used. Adobe is the choice for dirt-cheap construction. Anyone can do it and the adobes themselves don't necessarily need to be made in a form. They can be hand-patted into the desired shape and left to dry until ready to be mortared into place.

Earthbags, on the other hand, do not require as much time and attention as adobe. Since the bags act as a form, the mix is put directly into them right in place on the wall. Not as much moisture is necessary for earthbags as adobe. This is a distinct advantage where water is precious and scant. Earthbags cure in place on the wall, eliminating the down time spent waiting for the individual units to dry. Less time is spent handling the individual units, which allows more time for building. Even in the rain, work on an earthbag wall can continue without adversely affecting the outcome. Depending on the size, adobe can weigh as much as 40-50 pounds (17.8-22.2 kg) apiece. Between turning, moving, and lifting into place on the wall, each adobe is handled at least three or four times before it is ever in place.

Adobe is usually a specific ratio of clay to sand. It is often amended with straw or animal dung to provide strength, durability, decrease cracking, increase its insu-

SWSA

1.4: *Cleaning adobes at Rio Abajo Adobe Yard, Belen, New Mexico.*

SWSA

1.5: *The entire form box can be set in place using the Bobcat. Steel whalers keep forms true and plumb and resist ramming pressure.*

1.6: *Rammed earth wall after removal of forms.*

SWSA

lative value, and make it lighter. Earthbag doesn't require the specific ratios of clay to sand, and the addition of amendment materials is unnecessary as the bag itself compensates for a low quality earthen fill.

Rammed earth is another form of earth building that has been around for centuries and is used worldwide. Many kilometers of the Great Wall of China were made using rammed earth. Multi-storied office and apartment buildings in several European countries have been built using rammed earth, many of them in existence since the early 1900s. Rammed earth is currently enjoying a comeback in some of the industrialized nations such as Australia.

Rammed earth involves the construction of temporary forms that the earth is compacted into. These forms must be built strong enough to resist the pressure exerted on them from ramming (compacting) the earth into them. Traditionally, these forms are constructed of sections of lashed poles moved along the wall after it is compacted. Contemporary forms are complex and often require heavy equipment or extra labor to install, disassemble, and move (Fig. 1.5). The soil is also of a specific ratio of clay to sand with about ten percent moisture by weight added to the mix. In most modern rammed earth construction, a percentage of cement or asphalt emulsion is added to the earthen mix to help stabilize it, increase cohesion and compressive strength, and decrease the chance of erosion once the rammed earth wall is exposed.

While the optimum soil mix for both rammed earth and earthbag is similar, and both types of construction utilize compaction as the means of obtaining strength and durability, that is about where the similarity ends. Because the bags themselves act as the form for the earth, and because they stay within the walls, earthbag construction eliminates the need for heavy-duty wood and steel forms that are not very user-friendly for the one-time owner/builder. Since the forms are generally constructed of wood and steel, they tend to be rectilinear in nature, not allowing for the sweeping curves and bends that earthbag construction can readily yield, giving many more options to an earth builder (Fig. 1.6). While the soil mix for rammed earth is thought of as an optimum, earthbags permit a wider range of soil types. And just try making a dome using the rammed earth technique, something that earthbags excel at achieving.

Cob is a traditional English term for a style of earth building comprised of clay, sand, and copious amounts of long straw. Everybody loves cob.

It is particularly useful in wetter climates where the drying of adobes is difficult. England and Wales have some of the best examples of cob structures that have been in use for nearly five centuries (Fig. 1.7). Cob is also enjoying a resurgence in popularity in alternative architecture circles. Becky Bee and The Cob Cottage Company, both located in Oregon, have worked extensively with cob in the Northwestern United States. They have produced some very fine written material on the subject and offer many workshops nationwide on this type of construction. Consult the resource guide at the back of this book to find sources for more information on cob.

Simply stated, cob uses a combination of clay, sand, straw, and water to create stiff, bread loaf shaped "cobs" that are plopped in place on the wall and "knitted" into each other to create a consolidated mass. Like earthbag, cob can be formed into curvilinear shapes due to its malleability. Unlike earthbag, cob requires the use of straw, lots of straw. The straw works for cob the same way that steel reinforcing does for concrete. It gives the wall increased tensile strength, especially when the cobs are worked into one another with the use of the "cobber's thumb" or one's own hands and fingers (Fig. 1.8).

While building with earthbags can continue up the height of a wall unimpeded row after row, cob requires a certain amount of time to "set-up" before it can be continued higher. As a cob wall grows in height, the weight of the overlying cobs can begin to deform the lower courses of cob if they are still wet. The amount of cob that can be built up in one session without deforming is known as a "lift." Each lift must be allowed time to dry a little before the next lift is added to avoid this bulging deformation. The amount of time necessary is dependent on the moisture content

1.7: *Example of historic cob structure; The Trout Inn in the U.K.*

1.8: *Michelle Wiley sculpting a cob shed in her backyard in Moab, Utah.*

of each lift and the prevailing weather conditions. Earthbag building doesn't require any of this extra attention due to the nature of the bags themselves. They offer tensile strength sufficient to prevent deformation even if the soil mix in the bag has greater than the optimum moisture content. So the main advantages of earthbag over cob are: no straw needed, no waiting for a lift to set up, wider moisture parameters, and a less specific soil mix necessary.

Pressed block is a relatively recent type of earthen construction, especially when compared to the above forms of earth building. It is essentially the marriage of adobe and rammed earth. Using an optimum rammed earth mix of clay and sand, the moistened soil is compressed into a brick shape by a machine that can be either manual or automated. A common one used in many disadvantaged locales and encouraged by Habitat for Humanity is a manual pressed-block machine. Many Third World communities have been lifted out of oppressive poverty and homelessness through the introduction of this innovative device (Fig 1.9). The main advantage of earthbag over pressed block is the same as that over all the above-mentioned earth-building forms, the fact that earthbags do not require a specific soil mixture to work properly. Adobe, rammed earth, cob, and pressed block rely on a prescribed ratio of clay and sand, or clay, sand, and straw whose availability limits their use. The earthbag system can extend earthen architecture beyond these limitations by using a wider range of soils and,

when absolutely necessary, even dry sand — as could be the case for temporary disaster relief shelter.

Other Observations Concerning Earthbags

Tensile strength. Another advantage of earthbags is the tensile strength inherent in the woven poly tubing combined with the use of 4-point barbed wire. It's sort of a double-whammy of tensile vigor not evident in most other forms of earth construction. Rammed earth and even concrete need the addition of reinforcing rods to give them the strength necessary to keep from pulling apart when placed under opposing stresses. The combination of textile casing and barbed wire builds tensile strength into every row of an earthbag structure.

Flood Control. Earthbag architecture is not meant to be a substitute for other forms of earth building; it merely expands our options. One historic use of earthbags is in the control of devastating floods. Not only do sandbags hold back unruly floodwaters, they actually increase in strength after submersion in water. We had this lesson driven home to us when a flash flood raged through our hometown. Backyards became awash in silt-laden floodwater that poured unceremoniously through the door of our Honey House dome,

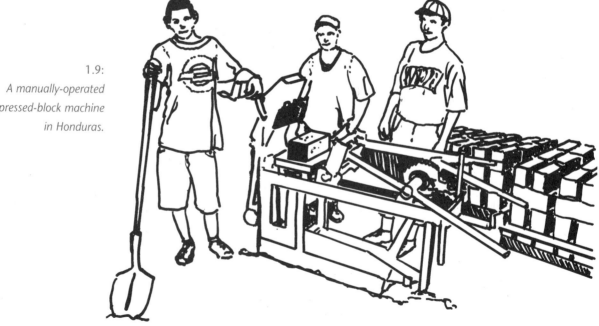

1.9:
A manually-operated pressed-block machine in Honduras.

leaving about ten inches (25 cm) of water behind. By the next morning, the water had percolated through our porous, unfinished earthen floor leaving a nice layer of thick, red mud as the only evidence of its presence. Other than dissolving some of the earth plaster from the walls at floor level, no damage was done. In fact, the bags that had been submerged eventually dried harder than they had been before. And the mud left behind looked great smeared on the walls!

Built-in Stabilizer. The textile form (bag!) encases the raw earth even when fully saturated. Really, the bag can be considered a "mechanical stabilizer" rather than a chemical stabilizer. In order to stabilize the soil in some forms of earth construction, a percentage of cement, or lime, or asphalt emulsion is added that chemically alters the composition of the earth making it resistant to water absorption. Earthbags, on the other hand, can utilize raw earth for the majority of the walls, even below ground, thanks to this mechanical stabilization. This translates to a wider range of soil options that extends earth construction into non-traditional earth building regions like the Bahamas, South Pacific, and a good portion of North America. While forests are dependent on specific climatic conditions to grow trees, some form of raw earth exists almost everywhere.

The Proof is in the Pudding

Nader Khalili has demonstrated the structural integrity of his non-stabilized (natural raw earth) earthbag domes. Under static load testing conditions simulating seismic, wind, and snow loads, the tests exceeded 1991 Uniform Building Code requirements by 200 percent. These tests were done at Cal-Earth — California Institute of Earth Art and Architecture — in Hesperia, CA., under the supervision of the ICBO (International Conference of Building Officials), monitored in conjunction with independent engineers of the Inland Engineering Corporation. No surface deflections were observed, and the simulated live load testing, done at a later date, continued beyond the agreed limits until the testing apparatus began to fail. The buildings could apparently withstand more

abuse than the equipment designed to test it! The earthbag system has been proven to withstand the ravages of fire, flooding, hurricanes, termites, and two natural earthquakes measuring over six and seven on the Richter scale. The earthbag system in conjunction with the design of *monolithic shapes* is the key to its structural integrity.

Thermal Performance

Every material in a building has an insulation value that can be described as an R-value. Most builders think of R-value as a description of the ability of a structure or material to resist heat loss. This is a steady state value that doesn't change regardless of the outside temperature variations that occur naturally on a daily and annual basis. So why does an earthbag structure (or any massive earthen building for that matter) with an R-value less than 0.25 per inch (2.5 cm) feel cool in the summer and warm in the winter? Because this R-value can also be expressed as the coefficient of heat transfer, or conductivity, or U-value, which is inversely proportional, that is $U = 1/R$. From this simple formula we can see that material with a high R-value will yield a low U-value. U-value (units of thermal radiation) measures a material's ability to store and transfer heat, rather than resist its loss. Earthen walls function as an absorbent mass that is able to store warmth and re-radiate it back into the living space as the mass cools. This temperature fluctuation is known as the "thermal flywheel effect."

The effect of the flywheel is a 12-hour delay in energy transfer from exterior to interior. This means that at the hottest time of the day the inside of an earthbag structure is at its coolest, while at the coolest time of the day the interior is at its warmest. Of course this thermal performance is regulated by many factors including the placement and condition of windows and doors, climatic zone, wall color, wall orientation, and particularly wall thickness. This twelve-hour delay is only possible in walls greater than 12 inches (30 cm) thick.

According to many scholars, building professionals, and environmental groups, earthen buildings

1.10: *Students working on Community Hogan on the Navajo Indian Reservation.*

currently house over one-third of the world's population, in climates as diverse as Asia, Europe, Africa, and the US with a strong resurgence in Australia. An earthen structure offers a level of comfort expressed by a long history of worldwide experience. Properly designed earthbag architecture encourages buried architecture, as it is sturdy, rot resistant, and resource convenient. Bermed and buried structures provide assisted protection from the elements. Berming this structure in a dry Arizona desert will keep it cool in the summer, while nestling it into a south-facing hillside with additional insulation will help keep it warm in a Vermont winter. The earth itself is nature's most reliable temperature regulator.

Cost Effectiveness

Materials for earthbag construction are in most cases inexpensive, abundant, and accessible. Grain bags and barbed wire are available throughout most of the world or can be imported for a fraction of the cost of cement, steel, and lumber. Dirt can be harvested on site or often hauled in for the cost of trucking. Developed countries have the advantage of mechanized gravel yards that produce vast quantities of "reject fines" from the by-product of road building materials. Gravel yards, bag manufactures, and agricultural supply co-ops become an earthbag builder's equivalent of the local hardware store. When we switched to earthen dome construction, we kissed our lumberyard bills goodbye.

Empowering Community

Earthbag construction utilizing the Flexible-Form Rammed Earth (FFRE) technique employs people instead of products (Fig. 1.10). The FFRE technique practices third world ingenuity, with an abundance of naturally occurring earth, coupled with a few high tech materials to result in a relatively low impact and

1.11: *Typical 1,000-year-old Anasazi structure, Hovenweep National Monument.*

embodied energy product. What one saves on materials supports people rather than corporations. The simplicity of the technique lends itself to owner/builder and sweat-equity housing endeavors and disaster relief efforts. Properly designed corbelled earthbag domes excel in structural resilience in the face of the most challenging of natural disasters. Does it really make sense to replace a tornado-ravaged tract house in Kansas with another tract house? An earthbag dome provides more security than most homeowner insurance policies could offer by building a house that is resistant to fire, rot, termites, earthquakes, hurricanes, and flood conditions.

Sustainability

Earthen architecture endures. That which endures sustains. Examples of early Pueblo earthen construction practices dating from 1250-1300 AD is evident throughout the Southwestern United States (Fig 1.11). The coursed adobe walls of Casa Grande in Southern Arizona, Castillo Ruins, Pot Creek Pueblo and Forked Lightning Pueblo in New Mexico, and the Nawthis site in central Utah, although eroded with centuries of neglect, still endure the ravages of time. In the rainy climate of Wales, the thick earthen cob-walled cottages protected under their thatched reed roofs boast some 300 to 500 hundred years of continual use. If we can build one ecologically friendly house in our lifetime that is habitable for 500 years, we will have contributed towards a sustainable society.

CHAPTER 2

Basic Materials for Earthbag Building

The Dirt

The dirt is the most fundamental element of earthbag construction. We strive for an optimal, rammed earth-soil ratio of approximately 30 percent clay to 70 percent sand. According to David Easton, in *The Rammed Earth House* (see Resource Guide), most of the world's oldest surviving rammed earth walls were constructed of this soil mix ratio. We like to use as close a ratio mix to this as possible for our own projects. This assigns the use of the bags as a temporary form until the rammed earth cures, rather than having to rely on the integrity of the bag itself to hold the earth in place over the lifetime of the wall. However, the earthbag system offers a wide range of successful exceptions to the ideal soil ratio, as we shall discover as we go on. First, let's acquaint ourselves with the components of an optimal earth building soil.

The Basic Components of Earth Building Soil

Clay plays the leading role in the performance of any traditional earthen wall building mix. Clay (according to Webster's dictionary) is a word derived from the Indo-European base *glei-, to stick together*. It is defined as, "a firm, fine-grained earth, plastic when wet, composed chiefly of hydrous aluminum silicate minerals. It is produced by the chemical decomposition of rock

2.1: *Wild-harvested clay lumps ready for pulverizing and screening.*

of a super fine particulate size." Clay is the glue that holds all the other particles of sand and gravel together, forming them into a solid conglomerate matrix. Clay is to a natural earthen wall what Portland cement is to concrete. Clay has an active, dynamic quality. When wet, clay is both sticky and slippery, and when dry, can be mistaken for fractured rock (Fig. 2.1). Sands and gravels, on the other hand, remain stable whether wet or dry.

One of the magical characteristics of clay is that it possesses a magnetic attraction that makes other ingredients want to stick to it. A good quality clay can be considered magnetically supercharged. Think of the times a wet, sticky mud has clung tenaciously to your shoes or the fenders of your car. Another of clay's magical traits can be seen under a microscope. On the microscopic level, clay particles resemble miniscule shingles that, when manipulated (by a tamper in our case), align themselves like fish scales that slip easily in between and around the coarser sand and gravel particles. This helps to tighten the fit within the matrix of the earth building soil, resembling a mini rock masonry wall on a microscopic level.

Not all clays are created alike, however. Clays vary in personality traits, some of which are more suitable for building than others. The best clays for wall building (and earth plasters) are of a relatively stable character. They swell minimally when wet and shrink minimally when dry. Good building clay will expand maybe one-half of its dry volume. Very expansive clays, like bentonite and montmorillonite, can swell 10-20 times their dry volume when wet. Typical clays that are appropriate for wall building are lateritic in nature (containing concentrations of iron oxides and iron hydroxides) and kaolinite. Expansive clay, like bentonite, is reserved for lining ponds and the buried faces of retaining walls or for sealing the first layer on a living roof or a buried dome.

Fortunately, it is not necessary to know the technical names of the various clays in order to build a wall. You can get a good feel for the quality of a clay simply by wetting it and playing with it in your hands. A suitable clay will feel tacky and want to stick to your skin. Highly expansive clay often has a slimy, almost gelatinous feel rather than feeling smooth yet sticky. Suitable clay will also feel plastic, and easily molds into shapes without cracking (Fig 2.2). For the purpose of earthbag wall building, we will be looking for soils with clay content of anywhere from 5 to 30 percent, with the balance made up of fine to coarse sands and gravels. Generally, soils with clay content over 30 percent are likely to be unstable, but only a field test of your proposed building soil will tell you if it is suitable for wall building.

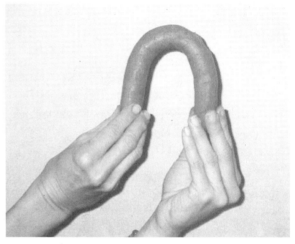

2.2: A plastic, stable quality clay can be molded with minimal cracking.

Silt is defined as pulverized rock dust, although its particle size is larger than that of clay yet smaller than that of fine sand. Silt is often present to a certain degree along with clay. It differs dramatically in behavior from clay as it is structurally inert. It mimics clay's powdery feel when dry, but has none of clay's active responses. It doesn't swell or get super sticky when wet. Too high a percentage of silt can weaken a wall-building soil.

Microscopically, silt appears more like little ball bearings than flat platelets like clay. It has a fine roly-poly feel that is designed to travel down rivers to be deposited as fertilizer along riparian corridors. All of nature has a purpose. Silt is just better for growing gardens than it is for building walls. Soils with an excessively high silt content should either be avoided or carefully amended with clay and sand before building with them. Building with soft, silty soil is like trying to build with talcum powder. In some cases, adding cement as a stabilizer aids in increasing binding and compression strength.

Sand is created from the disintegration of various types of rocks into loose gritty particles varying in size from as small as the eye can see to one-quarter-inch (0.6 cm), or so. Sand occurs naturally as a result of eons of erosion along seashores, riverbeds, and deserts where the earth's crust is exposed. Giant grinding machines at gravel yards can also artificially produce sand. Sand (and gravel) provides the bulk that gives an earthen wall compression strength and stability.

Sands have differing qualities, some of which are more desirable for wall building than others. As a rule of thumb, "well graded" (a term used to describe sand or soil that has a wide range of particle sizes in equal amounts), coarse, jagged edged sands provide more stable surfaces for our clay binder to adhere to. Jagged edged sand grains fit together more like a puzzle, helping them to lock into one another. Sand from granitic rock is usually sharp and angular, while sands from disintegrated sandstone are generally round and smooth.

Gravel is made of the same rock as sand only bigger. It is comprised of coarse jagged pieces of rock varying in size from one-quarter-inch pebbles (0.6 cm) up to two- or three-inch (5-7.5 cm) "lumps" or "cobbles." A well-graded soil containing a wide variety of sizes of sand and gravel up to one inch (2.5 cm) contributes to the structural integrity of an earthen wall. A blend of various sized sand and gravel fills all the voids and crannies in between the spaces created by the sand and gravel. Each particle of sand and gravel is coated with clay and glued into place. Sand and gravel are the aggregates in an earthen soil mix much the same as they are for a concrete mix. In a perfect earth-building world the soil right under our feet would be the optimal mix of 25-30 percent stable clay to 70-75 percent well-graded sand and gravel. We can dream, but in the meantime, let's do a jar test to sample the reality of our soil's character.

Determining Soil Ratios

The jar test is a simple layman method for determining the clay to sand ratio of a potential soil mix. Take a sample of the dirt from a shovel's depth avoiding any humus or organic debris. (Soil suitable for earth building must be free from topsoil containing organic matter and debris such as leaves, twigs and grasses to be able to fully compact. Organic matter will not bond properly with the earth and will lead to cavities later on as the debris continues to decompose.) Fill a Mason jar half full with the dirt and the rest with water. Shake it up; let it sit overnight or until clear. The coarse sands will sink to the bottom, then the smaller sands and finally the silt and clay will settle on top. You want to see distinctive layers. This will show the approximate ratios. To give a rough estimate, a fine top layer of about one-third to one-quarter the thickness of the entire contents can be considered a suitable soil mix. If there is little delineation between the soils, such as all sand/no clay or one murky glob, you may want to amend what you have with imported clay or coarse sand or help stabilize it with a percentage of cement or lime (more on stabilization in Chapter 4).

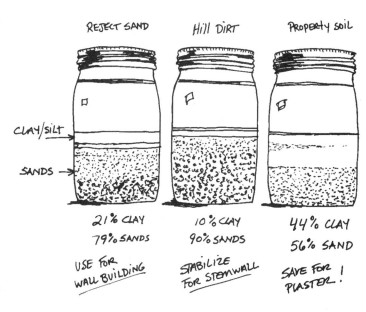

2.3: *The Jar Test. Three sample soils and their appropriate uses.*

Choose the best soil for the job. In some cases the choice of an earth building soil mix may depend on the climate. After a wall is built and standing for a few seasons some interesting observations can be made. Earthbag walls made with sandy soils are the most stable when they get wet. Cement/lime stucco over earthbags filled with a sandy soil will be less likely to crack over time than bags filled with a clayey soil. The richer a soil is in clay, the more it will shrink and expand in severe weather conditions. When building exposed garden walls in a wet climate, consider filling the bags with a coarse, well-draining soil and a lime/cement base plaster over stucco lath. Dry climates can take advantage of earthen and lime plasters over a broad variety of soil mixes as there is less chance of walls being affected by expansion and contraction.

Soils of varying ratios of clay and sand have unique qualities that can often be capitalized on just by designating them different roles. A soil sample with a high clay content may be reserved for an earthen plaster amended with straw. A sandy/gravelly soil is ideal for stabilizing with a percentage of lime or cement for a stem wall/foundation (Fig. 2.3).

Once we know our soil ratios from the jar test, we can go ahead and make a sample bag to observe the behavior of the soil as it dries and test its strength when cured. Seeing and feeling help us determine if we want to amend the soil with another soil higher in whatever may be lacking in this one, or give us the confidence that this soil is bombproof the way it is. If the soil is hopelessly inadequate for structural purposes, have no fear. Even the flimsiest of soils can still be used as non-load-bearing wall infill between a structural supporting post and beam system (refer to Chapter 5). Later on in this chapter, under "Soil Preparation and Moisture Content," we'll walk through how to make sample test bags.

Gravel Yards: Imported Soil. A convenient and common source for optimum to adequate building soil is often obtained at more developed gravel yards. This material is usually referred to as "reject sand" or "crusher fines." It is a waste by-product from the manufacture of the more expensive gravel and washed sand sold for concrete work. Reject sand is often the largest pile at the gravel yard and is usually priced dirt cheap. Our local reject sand has a ratio of approximately 20 percent clay to 80 percent sand/gravel. The primary expense is in delivery. For us it costs $58.75 to have 15 tons (13.6 metric tonnes) of reject sand delivered ($1.25 a ton for the dirt and $40.00 for the trucking). Another option for good wall building material is often called "road base." Road base usually has a higher ratio of gravel within its matrix, but still can be an excellent source for wall building especially as a candidate for cement stabilization for stem wall/ foundations.

Pay a visit to your local gravel yard before ordering a truckload. Take some buckets to collect soil samples in to bring home for making sample tests. You may find unexpected sources of soil that are suitable for your needs. This has largely been our experience when perusing gravel yards. Since a 600 square foot (58 square meters) structure can easily swallow up 50-80 tons (45-73 metric tonnes) of material, it is our preference to pay the extra cost of importing this clean, uniform, easy to dig (FQSS!), suitable clay/sand ratio mix for the sheer labor and time saving advantages. However, the beauty of earthbag building allows us the freedom to expand our soil options by using most types of soil available on site.

Exceptions to the Ultimate Clay/Sand Ratio

Steve Kemble and Carole Escott's Sand Castle on the Island of Rum Cay, in the Bahamas, is a wonderful example of the adaptability of earthbag architecture. All that was available to them was a mixture of coarse, crushed coral and sand so fine it bled the color and consistency of milk when wet. This material was obtained from the commercial dredging of a nearby marina. Because of the coarseness and size variety within the matrix of the fill material, it packed into a very solid block in spite of a clay content of zero percent (Fig 2.4).

A workshop in Wikieup, Arizona, introduced us to a similar situation of site-available coarse granitic sand that in spite of its low clay content (less than six percent) produced a strong compacted block of rammed earth. The sharp coarseness of this decom-posed granite fit like a jigsaw puzzle when tamped, locking all the grains together.

Marlene Wulf hand dug into a clay-rich slope of lateritic soil to build a bermed earthbag yurt in Georgia. (Fig. 2.5). The structures at Nader Khalili's school in Hesperia, California, are built of soil with only five percent clay content. Yet this coarse sandy mix has proven to endure shear and load bearing tests that have exceeded Uniform Building Code (UBC) standards by 200 percent.

Smooth surface sands from sandstone are generally considered weak soils for wall building. We've added cement to stabilize this type of earth and made it about as strong as a gingerbread cookie. Occasionally a situation arises where this kind of sand is our only option. Here's where the built-in flexible form allows us the opportunity to greatly expand our options from the ideal soil ratio. This is when, yes, we do rely on the integrity of the bag to a certain extent to stabilize the earth inside. In this case, we may consider build-ing an above ground post and beam infill, or a partially-buried round kiva style structure to support the brunt of the wall system (we would *not* consider building a dome with this weaker soil).

Soil Preparation and Moisture Content

Water plays a significant role in the preparation of the soil that will become the building blocks of our struc-ture. Although we coined the phrase flexible-form

2.4: *Doni harvesting crushed white coral in the Bahamas.*

rammed earth technique to describe the method to our madness, we have expanded our soil preparation recipes beyond what has been traditionally considered the ideal moisture content for a rammed earth soil. Before making a sample bag, we need to determine the ideal moisture content for the particular soil we are working with. All soils are unique and behave differently from each other. Each soil also behaves differently when prepared with differing amounts of water.

2.5: *Although labor intensive, this carefully excavated site did little to disturb the surrounding vegetation and provided the builder with the soil needed for her construction project.*

MARLENE WULF

The water content for rammed earth has traditionally been around ten to twelve percent. This percentage of moisture in an average suitable building soil feels fairly dry. It is damp enough to squeeze into a ball with your hand and hold together without showing any cracks (Fig. 2.6). A simple test is to moisten the soil and let it percolate evenly throughout the soil sample. Squeeze a sample of the earth in your hand. Next, hold the ball out at shoulder height and let it drop to the ground. If it shatters, that approximates what 10 percent moisture content feels and looks like.

2.6: *Squeeze a sample of the earth in your hand. There should be enough moisture that the soil compacts into a ball.*

This has long been considered the optimum moisture content for achieving thoroughly compacted rammed earth walls and compressed bricks. Ten percent moisture content allows a typical rammed earth soil mix to be pounded into a rock hard matrix and is hence considered the optimum moisture content. We too have followed the optimal moisture content practice in most of our projects.

However, we and fellow earthbag builders have made some discoveries contrary to the "optimum moisture content" as prescribed for rammed earth. We then discovered that our discoveries were previously discovered in laboratory tests conducted by FEB

Building Research Institute, at the University of Kassel, and published in the book, *Earth Construction Handbook*, by Gernot Minke. We found these test results fascinating for a couple of significant reasons.

Here's what we discovered. We can take a soil sample of an average quality earth mix of 17 percent clay, 15 percent silt, and 68 percent sand and gravel, and add about ten percent more water than the traditional ten percent moisture content prescribed for a rammed earth mix. The result produces a stronger yet less compacted finished block of earth. For those of you who are getting acquainted with building with earth for the first time, this may not seem like a big deal, but in the earth building trade, it flies in the face of a lot of people's preconception of what moisture content produces the strongest block of dirt.

Let's explore this a little further. Rammed earth is produced with low moisture and high compaction. When there is too much moisture in the mix, the earth will "jelly-up" rather than compact. The thinking has been that low moisture, high compaction makes a harder brick/block. Harder equals stronger, etc. What Minke is showing us is that the same soil with almost twice the ideal moisture content placed into a form and jiggled (or in the earthbag fashion, tamped from above with a hand tamper), produces a finished block with a higher compression strength than that of a ten percent moisture content rammed earth equivalent. What Minke is concluding is that the so-called optimum water content does not necessarily lead to the maximum compressive strength. On the contrary, the *workability* and *binding force* are the decisive parameters. His theory is that the extra moisture aids in activating the electromagnetic charge in the clay. This, accompanied by the vibrations from tamping, causes the clay platelets to settle into a denser, more structured pattern leading to increased binding power and, ultimately, increased compression strength.

We can take the same soil sample as above with lower moisture content and pound the pudding out of it, or we can increase the moisture content, "jiggle-tamp" it, and still get a strong block. What this means to us is less pounding (FQSS!). Tamping is hard work, and although we still have to tamp a moister mix to send good vibes through the earth, it is far less strenuous to jiggle-tamp a bag than to pound it into submission. Our personal discoveries were made through trial and error and dumb luck. *Weeper bag* or *bladder bag* are dirtbag terms we use when the soil is what we used to consider too moist, and excess moisture would weep through the woven strands of fabric when tamped. The extra moisture in the soil would resist compaction. Instead of pounding the bag down hard and flat, the tamper kind of bounced rather than smacked. The weeper bag would dry exceedingly hard, although thicker than its drier rammed earth neighbor, as if it hadn't been compacted as much.

We once left a five-gallon (18.75 liter) bucket of our favorite rammed earth mix out in the rain. It became as saturated as an adobe mix. We mixed it up and let it sit in the bucket until dry, and then dumped it out as a large consolidated block. It sat outside for two years, enduring storms and regular yard watering, and exhibited only the slightest bit of erosion. We have witnessed the same soil in a neglected earthbag made to the optimum 10 percent moisture specification (and pounded mercilessly), dissolve into the driveway in far less time. So now we consider the weeper bag as not such a sad sight to behold after all.

Our conclusion is that adapting the water content to suit the character of each soil mix is a decisive factor for preparing the soil for building. We are looking for a moisture content that will make the soil feel malleable and plastic without being gushy or soggy. The ball test can still apply as before, only now we are looking for a moisture content that will form a ball in our hands when we squeeze it; but when dropped from shoulder height, retains its shape, showing cracking and some deformation, rather than shattering into smithereens (Fig. 2.7).

Adjust the Moisture to Suit the Job

Personal preference also plays a role in deciding one's ideal mix. A drier mix produces a firmer wall to work on. Each row tamps down as firm as a sidewalk.

PROTECT FROM FREEZING

Earthbag construction is a seasonal activity. Need we say a frozen pile of dirt would be difficult to work with? Earthbag walls need frost-free weather to cure properly. Otherwise, nature will use her frost/thaw action to "cultivate" hard-packed earth back into fluffy soil. Once cured and protected from moisture invasion, earthbags are unaffected by freezing conditions.

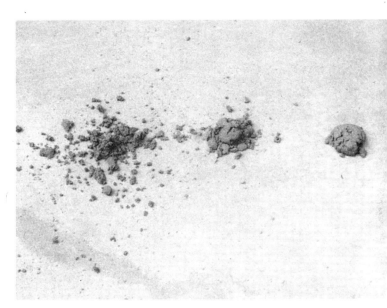

2.7: Three sample balls of soil dropped from shoulder height to the ground. The samples (left to right) show moisture contents varying from 10 to 20 percent.

If you have a big crew capable of constructing several feet of wall height in a day, a drier mix will be desirable. The moister the mix the more squishy the wall will feel until the earth sets up some. With a smaller crew completing two or so rows of bag work a day, a moister mix will make their job of tamping easier. You will have to be the judge of what feels best overall and meets the needs of your particular circumstances.

2.8: *Using a sprinkler to pre-moisten a pile of dirt in preparation for wall building.*

2.9: *In some cases where water is a precious resource or needs to be hauled to the building site, the earth can be flooded and held in check by tending little dams, allowing it to percolate overnight.*

Prepping soil (Fig. 2.8). Some soils need time to percolate in order for the water to distribute evenly throughout the pile. High clay soils require repeated watering to soften clumps as well as ample time to absorb and distribute the water evenly (sometimes days). Sandy soils percolate more quickly. They will need to be frequently refreshed with regular sprinklings (Fig. 2.9).

Make some sample test bags. To best understand soil types and moisture content, it's good to observe the results under working conditions, so let's fill and tamp some bags. When making test bags, try varying the percentage of water starting with the famous ten percent standard as a minimum reference point. For some soils ten percent may still be the best choice. For now, lets pre-moisten our test pile of dirt to about ten percent moisture.

Once the proper moisture content has been achieved (plan on a full day to a few days for this), fill some sample bags (refer to Chapter 3 for details on the art of *diddling* and *locking diddles* for making the most of your test bag). After filling, fold each bag shut and pin it closed with a nail. Lay the bags on the ground and tamp them thoroughly with a *full pounder* (see Chapter 3 for description of pounders and other tools). Let them cure for a week or more in warm, dry weather, protected from frost and rain. Thick rammed earth walls can take months to fully cure, but after a week or two in hot, dry weather, our test bags should feel nice and hard when thumped. Vary the moisture content in these test bags to get better acquainted with how they differ in texture while filling, how they differ while being tamped, and what the final dried results are.

After the bags are sufficiently cured, we test each one by kicking it, like a tire. We jump up and down on it and drive three-inch (7.5 cm) nails into the middle of it. If the soil is hard enough to hold nails and resist fracturing, it is usually a pretty good soil. If the soil is soft or shrunken, it will need to be avoided or amended or used as infill for a post and beam structure. We do these tests to determine which moisture ratio is best suited for this particular soil (for more scientific code-sanctioned tests concerning modulus of rupture and compression, we suggest consulting the New Mexico Uniform Building Code) (Fig. 2.10).

Our personal feeling is that earthbag construction should be tested as a dynamic system rather than an individual unit. It is the combination of all the ingredients — bags, tubes, soil, barbed wire, careful installation, and architectural design — that

2.10: (top) *This informal test demonstrates the weight of a 3/4-ton truck on top of a fully cured earthbag, resulting in no deformation whatsoever.*

2.11a: (top right) *The owners of this tall earthbag privacy wall, located on a busy intersection in town, woke up to find that the earthen plaster on one area of their wall had fallen off. The reason is shown in the next picture.*

2.11b: (lower right) *During the night, an unintentional "test" was conducted by an inebriated driver, which helped answer our questions about the impact resistance of an earthbag wall — the wall passed; the car failed.*

determine the overall strength of an Earthbag building (Fig. 2.11a & b).

Earth is a simple yet complex substance that you can work with intuitively as its merits become familiar. Experimentation is a big part of the earthen construction game. Once the test bags have dried, and the right soil mix and the suitable moisture content for the particular job has been chosen, the building crew is ready to go to work. A team of six to eight people can go through about 25 tons (22.5 metric tonnes) of easily accessible material in three days. Kept pre-moistened and protected with a tarp, it's ready for wall building throughout the week. If the building process is simple, the progress is quick.

2.12: *Bag ensemble (left to right): way-too-big; 100-lb. misprint; 50-lb. misprint; 50-lb. gusseted misprint; 50-lb. burlap.*

Bags and Tubes: The Flexible Form

The bags we use are the same kind of bags used most typically to package feed and grain (Fig. 2.12). The type and sizes we use most often are woven polypropylene *50-pound and 100-pound misprints* with a minimum ten-by-ten denier weave per square inch.

The companies that manufacture these bags sometimes have mistakes in the printing process that render them unsuitable to their clients. Rather than throw the bags away, they sell them at a considerably reduced cost. The 50-lb. misprint bags come in bales of 1000 bags and weigh about 120 pounds (53-54 kg) per bale. The more you buy the lower the price per bale. Prices for the 50-lb. bags average about 15-25 cents each, or from however much you're willing to pay to single-digit cents per bag for large orders (tens of thousands).

The average, empty "lay flat," 50-lb. bag (the term used by the manufacturers) measures approximately 17 inches (42.5 cm) wide by 30 inches (75 cm) long. When filled and tamped with moistened dirt we call it a *working* 50-lb. bag which tamps out to about 15 inches (37.5 cm) wide by 20 inches (50 cm) long and 5 inches (12.5 cm) thick, and weighs 90-100 pounds (40-45 kg). The typical lay flat 100-lb. bag measures 22 inches wide by 36 inches long (55 cm by 90 cm). A working 100-lb. bag tamps out to about 19 inches (47.5 cm) wide by 24 inches (60 cm) long and 6 inches (15 cm) thick, and weighs a hefty 180-200 pounds (80-90 kg). In general, whatever the lay-flat width of a bag is, it will become two- to three-inches (5-7.5 cm) narrower when filled and tamped with earth. These two sizes of bags are fairly standard in the US. Twenty-five pound bags are usually too small to be worthwhile for structural purposes. By the time they are filled and folded they lose almost half their length. In general, we have not bothered with bags smaller than the 50-lb. variety.

Larger bags, up to 24-inch lay-flat width (which we refer to as *way-too-big* bags), can also be purchased for special applications such as dormered windows in domes or a big fat stem wall over a rammed earth tire foundation.

This provides additional support for the openings, while giving the appearance of a wider wall. By using the wider bags or doubling up the 50-lb. bags, we can flesh out the depth of the windowsills for a nice deep seating area (Fig. 2.13).

It has recently come to our attention that bag manufacturers have been putting what they call a "non-skid" coating onto the polypropylene fabric. These treated bags and tubes should be avoided. The "non-skid" treatment reduces breathability of the fabric, keeping the earth from being able to dry out and effectively cure. When inquiring or purchasing bags, be sure that the bags you order do *not* have the "non-skid" treatment applied.

Gusseted woven polypropylene bags are slowly becoming available in misprints. Gusseted bags resemble the design of brown-paper grocery bags. When filled they have a four-sided rectangular bottom. They are like having manufactured pre-diddled bags (refer to Chapter 3). The innovative boxy shape aids in stacking large amounts of grain without shifting. Someday all feedbags will be replaced with this gusseted variety and diddling will become a lost art.

Burlap bags also come in misprints. Burlap bags will hold up exposed to the sun in desert climates for a year if kept up off the ground, and as long as their

2.13: The 100-lb. and way-too-big bags can also be used to surround the window and doorways in conjunction with the narrower 50-lb. bags/tubes for the walls.

SUSTAINABLE SYSTEMS SUPPORT (SSS).

seams have been sewn with a UV resistant thread. Otherwise, they will tend to split at the seams over time. In a moist climate they are inclined to rot. Stabilizing the earth inside them with a percentage of cement or lime could be an advantage if you want the look of a masonry wall to evolve as the bags decompose. Burlap bags come in similar dimensional sizes as the poly bags (Fig. 2.14). In the United States, they are priced considerably higher. The cost continues to escalate in the shipping, as they are heavier and bulkier than the poly bags. Contrary to popular assumption, natural earthen plaster has no discriminating preference for burlap fiber. Most burlap bags available in the US are treated with hydrocarbons. Some people have adverse physical reactions to the use of hydrocarbons including skin reactions, headaches, and respiratory ailments. Unfortunately, hydrocarbon treated bags are the type of burlap bag most commonly available to us in North America. Untreated burlap bags are called hydrocarbon free. The fabric is instead processed with food grade vegetable oil and remains odorless. Hydrocarbon free burlap bags require more detective work to locate but are definitely the non-toxic alternative. Perhaps as we evolve beyond our political biases, plant fibers such as hemp will be available for the manufacturing of feed bags. Bag manufacturers can be found on-line or in the Thomas register at your local library (refer to the Resource Guide at the back of this book).

The tubes, also called "long bags" or "continuous bags," are also made of woven polypropylene (Fig. 2.15). We use the flat weave variety rather than the style of tubes that are sewn on the bias. Tubes are what manufacturers make the feed bags from prior to the cut and sew process. Since they are not misprints the cost can be slightly higher per linear foot than the bags. The rolls can weigh as much as 400-600 lbs (181-272 kg) depending on the width of the material. They come on a standard 2,000-yard (1,829 m) roll, but sometimes the manufacturers are gracious enough to provide a 1,000-yard (914 m) roll. Tubes are available in all the same widths as bags. Tubes behave like the bags in that they lose two to three inches (2.5-3.75 cm) of their original lay-flat width when filled and

2.14: *Burlap bags have a nice organic look that can be appreciated during construction.*

TIP:

Burlap bags are floppy compared to polypropylene bags. As a result, they tend to slip easily out of the bag stand while being filled. To avoid this annoying habit, pre-soak the burlap bags to stiffen them up prior to placing on the bag stand and filling.

2.15: *Tubes are cut from a continuous bag on a roll.*

MARA CRANIC

2.16: *Tubes are the quintessential flexible form.*

THE ADVANTAGE TO KEEPING THE BAGS IN GOOD CONDITION ARE:

- In case of a flood or plumbing accident, the dirt will remain in the wall instead of a mud puddle on the floor.

- The bags are often easier to plaster over than the soil inside of them. An earthen wall likes to be covered with an earthen plaster that is similar in character. Sandy soil walls like a sandy soil plaster. A sandy soil plaster though, is not as resistant to erosion as a clay-rich plaster mix. Maintaining the health of the bag expands our plastering options.

- The bags provide tensile strength by giving the barbed wire something to grab onto. More bag, more grab.

tamped. Although 25-lb. bags are usually too small to use structurally, narrow 12-inch (30 cm) wide tubes (designed to become 25-lb. bags) make neat, narrow serpentine garden walls and slimmer walls for interior dividing walls inside earthbag structures.

Tubes excel for use in round, buried structures, free-form garden and retaining walls, and as a locking row over an arch (Fig. 2.16). Their extra length provides additional tensile strength for coiling the roof of a dome. They are speedier to lay than individual bags as long as you have a minimum crew of three people (refer to Chapter 3). Outside of the US, tubes also are available in burlap fabric and perhaps cotton. Our personal experience is limited to woven polypropylene tubes available in the US and Mexico.

Polypropylene bags are vulnerable to sun damage from UV exposure. They need to be thoroughly protected from sunlight until ready to use. Once you start building, it will take about three to four months of Utah summer sun to break them down to confetti. This can be a motivating factor to get the bag work done quickly with a good crew if maintaining the integrity of the bags is at all a priority. Most suitable rammed earth soils will set up and cure before the bags deteriorate. Even after the bags do break down a quality soil mix will remain intact. Still, there are advantages to keeping the bags in good condition.

While our little Honey House dome was still being finished a flash flood filled it, and all our neighbors' basements, with 10 inches (25 cm) of water. The base coat of the interior earthen plaster melted off the walls from 12 inches (30 cm) down. Since the floor had yet to be poured, the floodwater percolated into the ground.

The bags that were under water were soft enough to press a thumbprint into but not soggy. We supposed that under the extreme amount of compression from the weight of the walls above, the earth inside the bags were able to resist full saturation. As they dried out they returned to a super hard rammed earthbag again. The bag stabilized the raw earth even underwater. Had the bags been compromised by UV damage, it could have been a whole other story.

Nader Khalili had a similar experience in the sunken floor of one of his earthbag domes. Floodwater filled it about two feet (60 cm) deep for a period of two weeks. He documented the effects in conjunction with the local Hesperia building department and made the same observations we had. In essence, the bag is a mechanical stabilizer, as opposed to a chemical stabilizer such as cement, added to the earth. The bags provide us with a stabilizer as well as a form while still granting us the flexibility to build with raw earth in adverse conditions.

One way to protect the bag work during long periods of construction is to plaster as you go (refer to Chapter 13). Then, of course, there is always the method of simply covering the bag work with a cheap, black plastic tarp for temporary protection.

Another way to foil UV deterioration is by double bagging to prolong protection from the sun. Back filling exterior walls also limits their exposure to UV damage. It is possible to purchase woven poly bags with added UV stabilization or black woven poly bags designed for flood and erosion control. These will not be misprints, however, and will be priced accordingly. Polypropylene is one of the more stable plastics. It has no odor, and when fully protected from the sun has an indefinite life span. Indefinite, in this case, means we really don't know how long it lasts.

Barbed Wire: The Velcro Mortar

We use two strands of 4-point barbed wire as a *Velcro mortar* between every row of bags. This cinches the bags together and provides tensile strength that inhibits the walls from being pulled apart. Tensile strength is something that most earthen architecture lacks. This Velcro mortar, aided by the tensile quality from the woven polypropylene bags (and tubes, in particular), provides a ratio of tensile strength unique to earthbag construction. The Velcro mortar allows for the design of corbelled domed roofs, as the four-point barbed wire gives a sure grip that enables the bags or tubes to be stepped in every row until gradually the circle is enclosed.

Four-point barbed wire comes in primarily two sizes; 12½ gauge and 15½ gauge. The heavy 12½ gauge weighs about 80 lbs. (35.5 kg) per roll and the lighter 15½ gauge weighs about 50 lbs. (22 kg) per roll. Both come in ¼-mile lengths (80 rods or 0.4 km). We prefer to use the heavy gauge for monolithic structures, particularly for the corbelled domes. The light gauge is adequate for linear designs and freestanding garden walls. Four-point barbed wire can be obtained from fencing supply outfits, farm and ranch equipment warehouses, or special ordered from selected lumberyards.

Barbed wire weights (flat rocks or long bricks) are used for holding down the barbed wire as it is rolled out in place on the wall (Fig. 2.18). We have also made weights by filling quart-size milk cartons with concrete and a stick of rebar. They last indefinitely and won't break when dropped. Plastic one-half gallon milk jugs

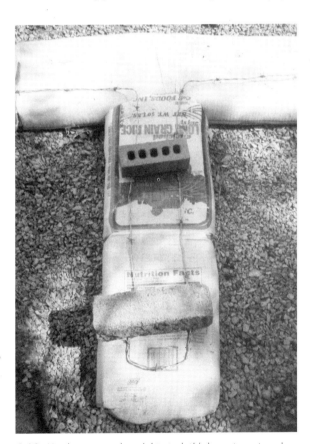

2.18: *Use long enough weights to hold down two strands of four point barbed wire per row at two- to three-foot intervals along the wall.*

2.19: *Buck stand converted into a barbed wire dispenser.*

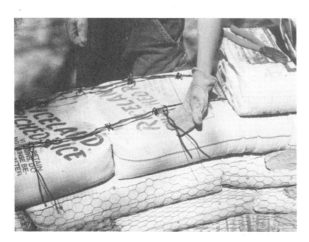

2.20: *Tie wire looped around barbed wire.*

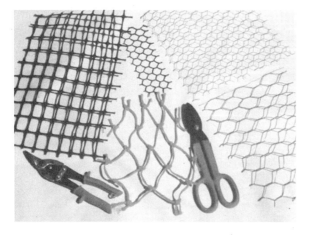

2.21: *Examples of a variety of plaster lath, also referred to as stucco mesh.*

filled with sand would also do the job, but would eventually break down from sun exposure. When we finally got tired of climbing up and down the walls to fetch rocks, bricks, and blocks, we created the multipurpose *suspended brick weights* (an FQSS innovation described in depth in Chapter 3).

A *barbed wire dispenser* can be made by placing a pipe through the roll of wire and supporting the pipe at either end with a simple stand made from wood or a stack of cinder blocks or fastened in between a couple of bales of straw. Or any other way you can think of that allows you to dole out a measured amount of the springy stuff. A *mobile wire dispenser* can be fashioned on top of a wheelbarrow or a manufactured version can be purchased from an agricultural farm or ranch supply catalog (Fig. 2.19).

Tie Wires

Tie wires provide an optional attachment source for the installation of chicken wire (stucco mesh) or a sturdy extruded plastic mesh substitute (Fig. 2.20). At the time of laying the barbed wire, one needs to decide whether cement/lime stucco, natural earth plaster, or earth plaster followed by lime plaster is going to be used as the finish coat. Clay-rich earth/straw plasters adhere directly onto the surface of the bags as tenaciously as they would to the cover of this book. Cement stucco requires chicken wire or a heavy gauge extruded plastic mesh (often used for concrete reinforcement and landscape erosion control). The main deciding factor between installing either a wire or a plastic mesh are weather conditions that would promote rusting of the metal wire variety in salt-air climates, a *living thatch* dome roof, or plastering work close to the ground where rain splash is likely to occur (Fig. 2.21).

Tie wires can be homemade cut sections from rolled 18-gauge wire or commercially available looped wire made for securing mesh fencing to metal stakes. Agricultural supply outfits and catalogues like Gempler's in the US offer a variety of inexpensive double loop steel and PVC-coated wire ties in packages of 100 eight-inch (20 cm) and twelve-inch (30 cm) lengths, with the twelve-inch (30 cm) variety being better suited for earthbag walls. These ties are shaped with a loop at both ends and are installed by folding the wire in half and wrapping the bent center around the barbed wire so that the two looped ends will protrude out beyond the wall. Commercial wire-ties (as they are referred to in the catalogues) are twisted tight with a manual or automatic wire-twisting tool. The manual one looks like a big crochet hook that is inserted through the two end loops and turned by hand. The automatic twisting tool has a spring-return action that twists the loops together with a pulling action rather than a twist of the wrist, and so is less tiring. Both tools are reasonably priced.

Tie wires are also used to secure electrical conduit and plumbing lines along interior walls (refer to Chapter 7). Tie wires are also used to anchor strawbales with exterior bamboo pinning cinched tight with extra long tie wires. (Look for illustrated details of this method in Chapter 17).

During construction we install long enough lengths of tie wire to project beyond the wall at least

two inches (5 cm). Secure the tie wires to the barbed wire every 12-24 inches (30-60 cm) every other row to provide an attachment source for the chicken wire later on. In addition, this provides an alternative fastening system for chicken wire other than nails. Most suitable rammed earth will hold a two-inch (5cm) or longer galvanized roofing nail for attaching stucco mesh after the walls have had sufficient time to cure. For added security and to avoid the potential of fracturing the earth, we may consider using the tie wires as an alternative attachment source. A single row of tie wires may be installed as a means of attaching a "weep screed hose" to create a "capillary break" between the plaster and the top of a stem wall (see Chapter 4 for more on this).

Arch Window and Door Forms

Although we use a flexible form for our walls we use a rigid form to make the empty spaces for our windows and doorways (Fig. 2.22). This is the only place that requires a temporary support system during construction (domed roofs are self-supporting). The *box forms*

2.22: Rigid form supporting door and window placement.

are leveled right on top of the wall. The bag work continues on either side of the form until the top is reached. The *arch forms* are then placed on top of the box form and leveled with wooden wedges inserted in between the arch and box forms. After the bag work of an arch is completed with the installation of the *keystone*

2.23: *Arch form being removed from the wall in the Bahamas.*

bags, the wedges are knocked out, and the arch form is dropped down and removed (Fig. 2.23 & 2.24).

Box and arch forms need to be ruggedly built to withstand the rigors of rammed earth construction. The thickness of the walls and whether the roof will be a corbelled dome dictate the depth of the forms. The forms need to be deep enough to accommodate the bag work as the rows are "stepped in" to create a corbelled domed roof. Three feet (90 cm) deep is often a versatile depth for dome building. Forms for linear walls only need to be a couple of inches deeper than the walls to prevent the bags from wrapping around the edges of them (or else you'll never get them out). In some cases, individual plywood paneling can be placed alongside a too narrow form to extend its depth. Add one inch (2.5 cm) more extra width and height to the forms to account for the rough openings, depending on the type of window and door systems being installed. Sculpted concrete, lime-stabilized earth, brick or stone windowsills need several inches of extra height to provide plenty of slope. Consider the window sizes and customize the forms accordingly or vise-versa.

Availability of materials and preferred style of the forms (open or solid) are also factors to consider. For

2.24: *After removal of forms. In curved walls, the columns in between the window openings take on an attractive trapezoidal shape.*

traditional header style doors and windows, the open *mine-shaft-style* door forms can be made using three-quarter-inch (1.875 cm) plywood or comparable siding material and four-by-four-inch (10x10 cm) or six-by-six-inch (15x15 cm) lumber (Fig. 2.25). Once the desired height of the opening is achieved, the disman-tled forms can become "lintels" (see Chapter 8).

Our favorite form system is a varying size set of *split box forms* and *solid arch forms* that can be used for dozens of structures (Fig. 2.26). One set of multiple size box and arch forms can be used to build an entire village of houses. They more than cover their initial costs in repeated use. Cinder blocks make handy forms for the rectilinear portions of the openings with wooden arch forms set on top. For the Bahamas Sand Castle project we had the delightful opportu-nity to borrow cinder blocks from our Bahamian friends who found the concept of "borrowing cement blocks to build a house" rather incredulous (Fig. 2.27).

To comply with FQSS approval, have *all* your window and door forms built for the structure before you begin construction. The structure is strongest built row by row with all of the forms in place, rather than pieced together in sections. It will save your san-ity, stamina and time to go ahead and have enough forms built for the entire project from the start.

It is conceivable to infill bags with dry sand as a non-wood substitute for box and window forms. These *sandbox bags* can take the place of wood or cement blocks in delineating the rectilinear portion of doors and windows. Use a plumb line to keep the out-side edges straight. Careful installation will be critical to maintain square. Allow extra room for error that can be filled in later with plaster around the window or doorjamb after construction. With a marking pen, denote where sandbox bags begin and regular earth-bags begin. Wrap chicken wire cradles around earthbags that butt up to sandbox bags to help delin-eate the difference. Remember to leave out the barbed wire on these sandbox bags or they won't come out later! (Fig 2.28).

2.25: *An open, mine-shaft style form allows easy access to the inside of a build-ing without climbing over the wall during construction.*

2.26: *Split box forms can be adjusted to accommodate various size openings.*

2.27: *Cinder blocks work well as temporary door forms.*

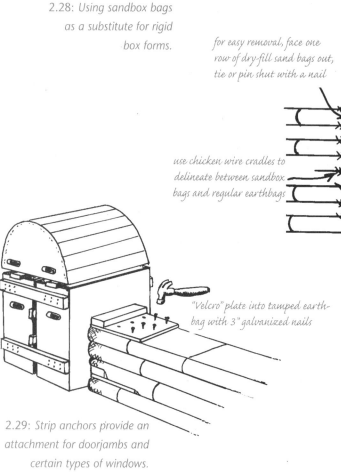

2.28: Using sandbox bags as a substitute for rigid box forms.

for easy removal, face one row of dry-fill sand bags out, tie or pin shut with a nail

wedge arch form on top of level board

use chicken wire cradles to delineate between sandbox bags and regular earthbags

install additional sand bags lengthwise

"Velcro" plate into tamped earthbag with 3" galvanized nails

2.29: Strip anchors provide an attachment for doorjambs and certain types of windows.

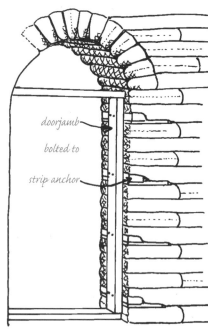

doorjamb bolted to strip anchor

2.30: Most doorjambs can be bolted to an adequate attachment surface that is provided by an average of four strip anchors spaced every three to four rows.

Velcro Plates

Doorjambs, shelf attachments, electrical boxes, intersecting stud frame walls, lintels, rafters, and extended eaves for domes, all need to attach to something that anchors them into an earthbag wall. *Velcro plates* are simply a flat wooden plate from one-half to one-inch (1.25-2.5 cm) in thickness, about twelve to sixteen inches (30-40 cm) long, cut to the approximate width of the wall and nailed into the bags. A *strip anchor* (a term used in adobe construction) allows for the later attachment of doorjambs after the forms are removed. A strip anchor is a length of two-by-four or two-by-six attached to a Velcro plate. It is then placed with the two-inch (5 cm) side flush against the box form and Velcroed (nailed) into the rammed earth bag below with two-and-one-half (6.25 cm) to three-inch (7.5 cm) long galvanized nails (Fig. 2.29). The bag work continues over the top of the strip anchors, incorporating them into the wall system during construction (Fig. 2.30). Windows can also be attached to strip anchors or can be shimmed and set into the walls with plaster alone.

A type of modified strip anchor is used for the placement of electrical boxes, lintels for rectilinear window and door frames, cabinetry, shelving, and anything that needs to be securely attached to the finished walls. A Velcro plate is used by itself to help distribute the weight of an eave or rafter across multiple bags.

The advantage of earthbag building is its minimal use of lumber. Although a finished earthbag structure can have a lot of Velcro plates and strip anchors throughout, it is still substantially less wood than in conventional construction (Fig. 2.31).

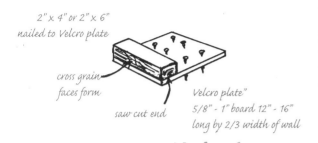

2" x 4" or 2" x 6"
nailed to Velcro plate

cross grain
faces form

saw cut end

Velcro plate"
5/8" - 1" board 12" - 16"
long by 2/3 width of wall

Note: if wide boards are unavailable, use two narrower boards side by side — pallets are an excellent source for strip anchor materials

2.31: Anatomy of a strip anchor.

Scabs

A "scab" is a Velcro plate used to connect a buttress into a wall or connect two rows of bags stacked side by side in a situation where this is more efficient than to stagger the bags in a mason-style running bond (Fig. 2.32).

If two-by-four lumber proves hard to scavenge, substitutions can be made with one-inch (2.5 cm) dimensional lumber commonly found in discarded pallets. For the strip anchor as well as the Velcro plate, the cross-grain of the wood is stronger to screw into than the saw cut ends.

Have several precut Velcro plates and scrap two-by material on hand when you start a project so that when you come to a point in the construction where a strip anchor or Velcro plate is needed, the work won't have to wait while you measure and cut these necessary items. We will learn more about Velcro plates and where to use them throughout this book (Fig. 2.33).

2.32: As a buttress gets shorter near the top of a wall, it is simpler to interlock the bags with a "scab," rather than try to make two dinky bags fit.

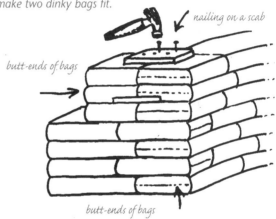

nailing on a scab

butt-ends of bags

butt-ends of bags

2.33: An excellent source of scrap lumber -conventional wood-frame construction sites.

2.34: *Cradles provide the underside of arches with an extra grippy surface for the later application of plaster.*

Cradles

Cradles are cut sections of chicken wire, extruded plastic mesh, woven split bamboo reed, or any suitable substitute that can be used to cradle the underside of each *fan-bag* (the bags that surround the arch forms) during construction. We still use cradles even when we intend to apply an earthen plaster, as this is the one place where the bags have conformed to such a smooth surface that the plaster needs something extra to key into. Cradles also work well installed around the bags that go up against the box forms. Cradles can be cut the exact width of the wall or extended as an anchor for sculpting an adobe relief pattern on the interior and exterior surfaces of the arches. This adds a dimension of artistic practicality for the design of drip edges and rain gutter systems (Fig. 2.34). See Chapter 6 for cradle installation.

CHAPTER 3

Tools, Tricks, and Terminology

Prior Preparation, Patience, Practice, and Perseverance Promote Preferred Performance (Fig. 3.1)

Just as it's easier to drive a nail with a hammer than a rock, a bag stand and slider assist in the ease of earthbag construction. To comply with the FQSS standard, we have developed a few specialized site-built tools and adopted techniques and a language that enhance the precision, quality, understanding, and enjoyment of earthbag building.

No matter how we build, building a house is a lot of work. Building a house is a process. What we learn from the process will be reflected in the product. The process proceeds smoothly when we pay attention to details, and attention to details begins with *prior preparation.*

Any professional builder or artisan will tell us that 75 percent of building time is spent preparing for the actual construction. That's why it is imperative to find joy in the process as well as the product. We spend most of our time and energy involved in the process, so let's make the most of it. In this modern world of instant gratification, the reality of a full-blown construction project can be daunting. Maybe there should be some sort of Home Builder's Anonymous organization that first timers could attend — kind of a 12-step program for acquiring a Zen philosophy toward building. The mantra

would be: prior preparation, patience, practice, and perseverance promote preferred performance.

Whenever we've tried to cut corners, we ended up having to backtrack, undo, and redo. It's the price we paid for our impatience. It is far cheaper to pay attention up front than to pay later by doing it all over again. Living with results that make us feel good every time we look at them is far more satisfying than wishing we'd taken the time to do a nice job. And if we stick with it, we will be rewarded in the end with a

3.1: *Tools of the dirtbag trade.*

33

work of beauty and a wealth of gained knowledge. Impatience, whining, and complaining are exhausting. Another way to think of building is like having a first baby. The more we are prepared to take care of another human being, the more fun we'll have.

Evolution of the Bag Stand

The bag stand holds the bag open in place on the wall freeing up both hands while you fill it. For us, the bag stand has evolved from the open-ended sheetrock bucket to our current favorite: a collapsible, lightweight, weld-free metal bag stand. We discovered the collapsible bag stand idea on a remote island in the Bahamas while scavenging for materials with which to build forms and tools. Rummaging through an abandoned, hurricane ravaged restaurant, Doni found a plastic food-serving tray stand. Turned upside down and trimmed, it was a perfect fit for the 100-lb. bags we were using to build Carol Escott's and Steve Kemble's Sand Castle on Rum Cay. Now we make our own simple, weld-free collapsible bag stand from common one-half-inch (1.25 cm) or three-quarter-inch (1.875 cm) flat-stock steel. A drill is the only tool required for drilling the pivot holes for the nut and bolt to go through.

3.2: *Evolutionary variety of bag stands.*
Left to right: collapsible wood; welded rigid metal; and our favorite, weld-free collapsible.

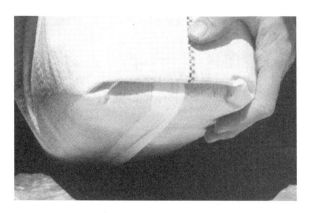

3.3: *A perfectly "diddled" bag.*

Along the evolutionary path, we developed the rigid, welded, metal bag stand, which requires some skill and access to welding equipment. Wooden bag stands are another option, but end up being more bulky and less sturdy in the long run (Fig. 3.2). (Refer to Appendix A for directions on building both types of metal bag stands).

Diddling

We like to give credit where credit is due. The first little experimental dome we worked on was a collaborative effort — a big party in one weekend. We were all occupied filling and flopping bags around when Doni looked up at Chaz, who was bent over intently fiddling with the bottom corners of a bag. "Chaz, what're you doing?" With a gleam in his eye, Chaz responded, "I'm diddling the bag."

What does diddling do? Diddling inverts the corners of the bag in a way that resembles a square-bottom brown-paper grocery bag (Fig. 3.3). Most bag work we'd been introduced to had a kind of primitive or downright sloppy appearance. This seemed OK, until it came time to plaster. All these bulging soft spots suddenly stuck out like sore thumbs. It took gobs more plaster to build the surrounding wall out to meet the bulges. Even when we went with the contours, the bulges posed still another problem. They are soft. The dirt hides in the corners avoiding compaction. The corners are floppy like rabbit's ears, making it hard for the plaster to stick. Even if you are going to cover the surface with chicken wire, the plaster will bond better to a firm subsurface (Fig. 3.4).

Diddled dirt bags make nice tight vertical seams where they meet, producing a neat, uniform appearance. Every part of the bag is hard. The Navajos preferred the term "tucking in the rabbit ears," as the word diddling is difficult to translate. In a chapter of *Alternative Construction*, edited by Lynne Elizabeth and Cassandra Adams, one of the authors referred to diddling as "invaginating" the bags. That's a little too clinical for us, but whatever you want to call it, the results are still FQSS (Fig. 3.5). If a diddled bag comes un-diddled during installation, finish laying it down and re-diddle it by shoving a pair of pliers into the corner or hammering a dowel to re-invert the corner (a.k.a.: "dimpling an undone diddle").

3.4: *Our early bag work resembled stacks of feed sacks with their soft corners bulging out.*

Locking the Diddles

Where an end bag is going to be exposed, like at the end of a buttress or a corner, we like to *lock the diddles* so they will remain intact when we tamp the row from above (see Chapter 6 for the complete directions on how to lock the diddle).

Pre-Diddled Bags (Fig. 3.6)

Gusseted bags are factory *pre-diddled* grain bags. It's funny how things work out. Never underestimate the power of good public relations. Kaki sent photos of the Honey House to our bag broker and he sent us some gusseted bags to play with. He said, "The

3.5: *Results of a perfectly diddled bag meet FQSS approval.*

3.6: *Inside-out gusseted bag on top of a diddled and locked bag. With the introduction of gusseted bags, diddling may become a lost art.*

grain bag industry has been experimenting with 'gusseted' square bottom bags to reduce stacked bags from shifting on pallets. It might work better for you guys, too."

Of course we save time spent from all that diddling, but we still hand pack the corners and bottoms of the bags to firm them up for a nice tight fit. We also like to turn the gusseted bags inside out for any exposed ends, like a buttress or corner. The bottom fabric is tight and smooth, a great surface for plaster. (For more on gusseted bags, refer to Chapter 2).

The Humble #10 Can (Fig. 3.8)

We use cans as hand shovels for scooping dirt out of wheelbarrows and passing it along to be dumped into a bag stood up on the wall. In terms of canned goods, they hold about three quarters of a gallon (2.8 liters). One can of dirt is approximately equivalent to one shovel of dirt. In our years of bag building, we have yet to discover a more effective way to move tons of dirt onto the walls than by hand with a can. A shovel tends to be awkward in that the handle swings around in someone's way, and it is harder to find a place to set it on the wall when you're ready to fold and lay the bag down (Fig. 3.9).

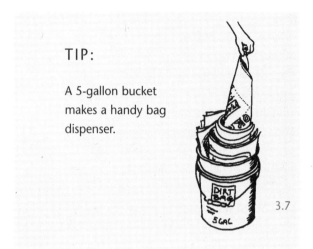

3.7

Can Tossing

As the walls grow taller, we toss the cans of dirt up to our partner on the wall or scaffolding. This may sound like an uncomfortable or awkward way to get the dirt up onto a wall, but compared to lifting a 100- or 200-pound (45-90 kg) bag onto the wall, an eight-pound can of dirt is beautiful in its simplicity. In fact, it's more like a cooperative non-competitive sport.

3.9: *Passing the can.*

3.8: *The large restaurant-size tomato can, coffee can, etc., are called "#10 cans."*

If you have ever pitched a softball, taken an under-handed shot with a basketball, rolled a bowling ball, golfed, or played catch with a four year old, you're a natural born can tosser. Though each person has their own style, the basic technique is the same. Grasp the can in both hands, under the can or on the sides close to the bottom. Some people keep their dominant hand on the bottom and roll the can upwards off the hand, using their other hand to guide the direction and control any tumbling effect. As with anything else it takes a little practice, but within a few tosses your cans will be ending up where you want and with little dirt spillage. Not that dirt spillage should ever present a problem (Fig. 3.10).

As you get used to tossing, catching, scooping, and dumping, a rhythm develops within the crew that is comfortable, enjoyable, and only interrupted by folding and laying a bag in place.

As a can receiver on the wall, besides catching the upcoming can, you also must coordinate dropping the empty can back into the wheelbarrow ready for the tosser while they toss a full one to you. When it all works smoothly, the upcoming full can literally floats into your waiting hands and all you need to do then is dump and drop. Of course, there is nothing wrong with handing your partner a can of dirt when it feels more convenient and comfortable to do so.

The humble #10 can is an indispensable friend to the earthbag builder. It easily enables the transformation of a mountain of dirt into a structure of beauty and solidity. It's used as a measuring device and a convenient water container for keeping your hands wet for plastering. And it makes a nice drumming sound for those musical breaks. Most restaurants are an excellent source for used #10 cans.

Sliders

After laying our Velcro mortar (4-point barbed wire), a *slider* is placed on the wall and the bag stand is set up on top of the slider. This allows the bag to be slid into the desired position on top of the barbed wire without getting stuck to it. Make sure to place the slider where you want the bag to be filled. The bag

will still be heavy but relatively easy to maneuver on the slider until it is laid down onto the barbed wire. (Fig. 3.11).

3.10: *Keep your eye on where you want the can to go, give it a little windup, and loft it upwards.*

3.11: *Filling a bag on top of a slider.*

Sliders can be made from any heavy gauge metal sheathing, from used galvanized heating ducts to hammered metal roofing. Useful sizes range from 16 inches (40 cm) by 18 inches (45 cm), or smaller or larger depending on the size of the bags being used. A variety of sizes are handy for tight spaces next to forms or for closing in a gap where two teams of wall builders are destined to meet.

3.12: *The row is tamped from above until the dirt goes from making a dull thud sound to a distinctive ringing sound.*

3.13: *Full pounders and quarter pounders.*

Tampers

Tampers are the hand ramming tools that turn moist, fluffy soil in a bag into a hard, compacted block. We use two sizes of tampers for two specific purposes. *Full pounders* are used to compact the bags or tubes after an entire row has been laid. Usually three to four passes are enough to perform the task of adequate compaction (Fig. 3.12).

The other type of tamper we use is smaller and called a *quarter pounder*. The quarter pounder is used primarily *inside* the bags to pre-tamp and/or shape the bag into a specific conformation. We use it to make a *hard-ass* bag for extra firm exposed end bags (buttresses) or for shaping a *fan bag* being built around an arch opening. The quarter pounder is also specific for tamping the *keystone* bags to lock an arch in place. A quarter pounder is narrower at the bottom (instead of wider at the bottom like a full pounder). We use the large size yogurt container (open, wide side up) to get a wedge shape ideal for tamping hard-ass bags and custom fan bags. We make two types of quarter pounders, one with a long handle that is comfortable to use while standing, and one with a short handle that is easy to use while sitting (Fig. 3.13).

Why do we make our own tampers? First, they are cheap to make (refer to Appendix A for practical plans and directions). Second, a full pounder is the most comfortable hand tamper we have ever used. They are round so there are no corners to gouge the bag, and we can control the weight by the size of the form we pour the concrete into. We have found the ideal weight to be about 12 pounds (5.5 kg). An upside-down plastic planter pot with a six-inch (15 cm) diameter filled up seven inches (17.5 cm) with concrete weighs about 12 pounds. Anything heavier will wear you out. Anything lighter or much wider at the bottom does an inadequate compaction job.

Hard-assing

Hard-assing is a technique (trick) we use to make a bag with an extra firm bottom. It adds extra compaction to any exposed bag at the end of a wall, buttress, or corner. We also hard-ass the bags up against the window and door forms for extra compaction (except for a bag that goes on top of a strip anchor). The extra firm bottom helps the end of the bag maintain its shape when later tamped from above. Without hard-assing, the end bags would slump when tamped from above (Fig. 3.14).

Feel free to change our lingo to suit your own comfort level. One workshop with high school students got a kick out of reminding each other to hard-ass the bag. Another workshop with a Christian church group coined the phrase "hard-bottom." One little girl participant offered to "do that thing to the bag that I'm not allowed to say the word for."

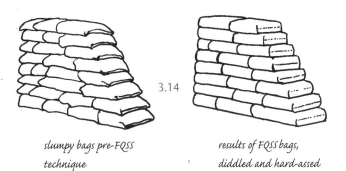

slumpy bags pre-FQSS technique

results of FQSS bags, diddled and hard-assed

Bag Whacking

Some people like to smack the sides of the bags during construction to flatten their profile. The logic is that flattening the surface of the wall will reduce the depth of the voids to fill with plaster later on. Although there may be some merit to this practice when one is intending to swaddle the entire structure in stucco mesh, the downside is that bag whacking tends to loosen the nice tight fit that was achieved while compacting the soil from above. In other words, the sides of the bag have been stretched tight from tamping. Side whacking tends to loosen the tension of the fabric, resulting in a baggy bag. Baggy bags are hard to plaster. Plaster likes a firm bag to bond with.

Suspended Brick Weights

We started out like everybody else using the most primitive available strategies for building earthbag walls. We used flat rocks or bricks to hold the springy barbed wire down while laying the next row of bags. This works fine for building a low garden wall. As the wall gets higher, though, it becomes a pain in the butt to have to keep heaving the bricks on and off the wall. So, in keeping with FQSS, we devised a technique that solved the brick-heaving dilemma and turned out to serve another purpose.

3.15: *Left to right: Loaded suspended brick weight; homemade wire hanger clip; store-bought spring clip.*

We bought a roll of polypropylene bailing twine used for tying bales of straw or hay. We tied a metal spring clip onto one end of 50 feet (15 meters) of twine and wound the twine around a brick. The loose end of the twine is then tied to the first row of barbed wire and clipped with the spring clip between the brick and the barbed wire so that the twine doesn't unravel and the brick will hang freely (Fig. 3.15).

After the next row of bags is laid and tamped, the next two rows of barbed wire goes on. We then swing the weighted twine over the top of the wall suspending the brick on the other side. The weighted twine holds down the wire and the brick is out of the way, yet easily accessible to reel in from atop the wall, as it grows taller. The twine is unwound from the brick as it is woven back and forth between each row

3.16: *Interweaving suspended brick weights holds down barbed wire and weaves each row together.*

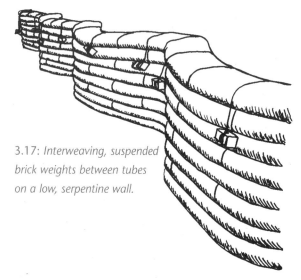

3.17: *Interweaving, suspended brick weights between tubes on a low, serpentine wall.*

and clipped into a locked position at each desired length (Fig. 3.16).

The initial investment may be more; we paid between 35 and 46 cents for new bricks, less than one dollar for each clip, and about $25 for one 1,000 foot (300 meters) roll of bailing twine. We use one suspended brick set-up about every two feet (60 cm) on the wall with a few extra for the ends of buttresses and corners. So, for 30 feet (9 meters) of wall we need at least 15 setups, or 50 brick set-ups for 100 feet (30 meters) of wall, etc. The additional money and time spent prepping the weights more than pays for itself in fewer backaches and less down time. Inexpensive substitutes for bricks can be, of course, free salvaged bricks, plastic one-half gallon (or 2-liter) milk jugs filled with sand and homemade clips made from wire coat hangers.

We mentioned another purpose the suspended brick weight serves. A low (up to four feet (1.2 m) high) serpentine curved wall built with tubes (see Tube Chutes in this chapter) would not necessarily require barbed wire. The solidity of the tubes laid in an S-shaped curve inhibits shifting. We feel we can limit the use of the barbed wire to the first row as a "tie-on" for the suspended brick weights. Tying all the rows of tubes together by weaving the suspended brick weights between each course helps to remind the rows they are one (Fig 3.17).

Fan Bags (Fig 3.18)

A *fan bag* is a specific type of earthbag. It is specific in its shape, size, and function. It is used exclusively to form the opening of an arch. Because it is strictly used to conform to the shape of an arch form, the bag is treated differently from a bag in the wall itself. It is always filled and tamped 12 inches (30 cm) in height. A bag in the wall will be tamped flat while a fan bag is hand-shaped and tamped in the shape of a wedge with the narrow end down, or rather, against the arch form. This is done to accommodate the curve of the arch. The inside circumference of a curve is always

3.18: *The fan of an arch.*

shorter than the outside circumference. Therefore, the fan bags must be wider at the top ends than at the bottom. They got the name fan bags from the way they resemble an open, folding hand fan.

The Wedge Box (Fig 3.19)

The *wedge box* is a special form for pre-tamping the fan bags that surround the arch form. The wedge box is designed to pre-shape a proper wedge shape to accommodate any size Roman arch by simply adding or reducing the number of fan bags used. The wedge box holds a bag that is filled and tamped 12 inches (30 cm) high. If you fill a fan bag in the wedge box on top of the wall, the wedge box then can be opened to slide the pre-tamped fan bag into place around the arch form. You can use the wedge box on the ground, as well, and then the pre-tamped fan bag can be handed up to a worker on the wall (Fig. 3.20, 3.21 & 3.22). (The directions for making your own wedge box can be found in Appendix A).

3.19: *The wedge box is a form for creating uniform fan bags.*

3.20: *Using the wedge box to form fan bags.*

3.21:
The ends of the wedge box are removed and the sides drop down for easy removal of the pre-tamped fan bag.

3.22: *The fan bags only weigh about 40 pounds (18 kg), the approximate weight of an average American-made adobe brick.*

Keystone Bags (Fig 3.23)

Keystone bags function in a bag arch the same way that a key *stone* does in a stone arch. Two to three bags are used as the keystone, depending on the type of arch being built. The purpose of the keystone bag is to force pressure down and out that is met with resistance by the walls on either side of the arch. (This is referred to as the buttressing of an arch.) This resistance directs the compression forces above the arch to the side of the opening and down to the ground. The gravitational forces of the earth embrace an arch. The dynamics of an arch are truly one of nature's magical feats of engineering. (For detailed instructions on keystone bag placement, refer to Chapter 6).

Tube Chutes (Fig. 3.24)

Bags use a stand. *Tubes* (continuous bag on a roll) use a *chute*. A tube chute can be made from a 20-inch (50 cm) long piece of sheet metal duct taped into a tube or a 20-inch (50 cm) section of sturdy cardboard, called a Sono tube, used as a form for pouring concrete foun-

3.24: *A tube chute acts as a sleeve to scrunch a length of tube onto, as well as a funnel that the dirt is fed into.*

3.23: *It is stronger to have too narrow of a gap for the keystones to fit into than one too wide.*

dation piers. Typically, we use an eight-inch (20 cm) diameter Sono tube for a 16-inch (40 cm) lay-flat width woven polypropylene tube. The poly tube is held onto the chute with a Bungee cord. You can make the tube chute any length that is comfortable for you. Use different diameter chutes for a variety of tube widths.

Loading a Tube Chute (Fig. 3.25-3.28)

A 15- to 20-foot (4.5-6 m) length of 16-17-inch (40-42.5 cm) lay-flat width of tubing can be loaded onto an 8-inch (20 cm) diameter by 20-inch (50 cm) long chute. If the diameter of the chute is too wide, it's hard to scrunch the tube on.

3.25, 3.26, and 3.27: *Loading a tube chute.*

3.25: Measure and cut the length of tube you want to lay and add an extra two to three feet (60-90 cm) to tie off the ends. Drape the tube over the top of a sawhorse or chair and pull the tube over the top of the chute.

3.26: Lift the chute and clasp a Bungee cord around the tube and chute. Kaki likes to fold the fabric down like a cuff and then clasp it, as the fabric tends to unravel.

3.27: Put the tube chute back on the ground and scrunch the remaining tube down over the chute like a stocking.

3.28: When you get to the end, pull it through the opening of the chute and tie the bottom closed with a string or strand from the unraveled tubing. Your tube chute is now loaded.

Katherine Huntress tying off a variety of different sizes of loaded tube chutes.

Laying a Coil

To lay a tube or, to be more romantic, lay a coil, hold the chute in a way that you can control the slow release of the tubing as your partner is filling it. Did we mention that it works better as a two-person job? That's because it really works best as a three-person job. For the most efficient use of your body and time, laying coils is most easily managed with a three-person team (Fig. 3.29).

At a low wall level, two people and several strategically placed wheelbarrows of dirt can manage just fine. For taller walls and domes, three people are needed. Oh, you can fill a tube by yourself by doing the job of all three people - tossing in dirt, jiggling it down the tube over and over, back and forth, climbing up and down the wall as it gets taller. Personally, if I had to do the work by myself, I'd use bags so I could free my

SUSTAINABLE SYSTEMS SUPPORT (SSS)

3.30: *Tube work is integrated with the fan bags forming an arched opening, rather than taking the place of fan bags.*

3.29: *The person holding the chute is the "walking bag stand," while the "tube loader" fills the chute and the third person delivers dirt and feeds cans to the loader.*

hands or I would wait until I had the money to hire a couple of people to help me. The beauty of this technique is that it doesn't take a college graduate to hand you a can of dirt. Spread the knowledge by paying for the help of unskilled labor to learn a new skill.

Tips for Tubes

Even when we are building with tubes, we prefer to integrate tubes with bags around the door and window openings, for two reasons. A bag always has the factory end-seam facing the form, whereas tubes are open-ended, so any time a tube is tied off at a door form the open end will be exposed. Next, the bags against the form get hard-assed on the inside before they are installed, as well as a second ramming from on top when the whole row is tamped. This double whammy treatment ensures that the earth around the more exposed openings is extra strong (Fig. 3.30). Lastly, the full tamped width of the bottom of a bag fills out a chicken wire *cradle* nice and evenly. FQSS! (For more on chicken wire *cradles* see Chapter 2).

Hard-Assing the Butt-End of a Tube

Overfill the tube two inches (5 cm) beyond the end of the wall. Twist the end of the tube and fold it back. Set a weight on the fabric to hold it out of the way. With a quarter pounder, tamp the end of the tube back in two inches (5 cm) shorter than the length of the finished wall (Fig. 3.31).

Re-twist the fabric extra tight to take up the slack. Lift up the end of the tube and tuck the twisted fabric underneath itself (Fig. 3.32).

3.31

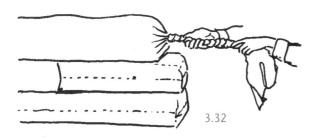

3.32

3.34

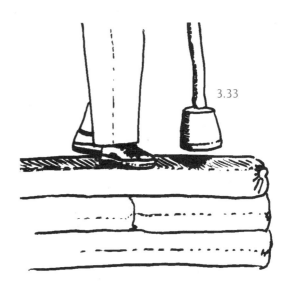

3.33

Tamp the whole row working from the middle of the tube out towards the ends. The tube will tamp out flush with the finished row below (Fig. 3.33 & 3.34).

Teamwork

As the wall gets taller, earthbag building works best as a collaborative effort. If you are only building a low garden wall with bags, you have both hands free to work by yourself. With several wheelbarrows filled with dirt and arranged along the wall, a single person can build alone. The progress escalates considerably,

3.35: A five- to seven-person team built the 9'6" tall walls for this 750 sq. ft. Bureau of Land Management Ranger Station in eight days.

however, when the work is shared. Taking turns keeping the wheelbarrows full and tossing cans of dirt up to a partner on the wall keeps the momentum going (Fig. 3.35).

We like to use a three-person team for low walls and five to seven people for taller walls. An odd number of workers keep the teams of two busy tossing and filling while the extra person keeps the wheelbarrows filled. For a really big project, it can keep two people busy just to maintain full wheelbarrows for all the wall builders.

When working with a crew, make sure that you first do a demonstration on how you want the bags filled so that everyone's work is consistent with everyone else's. Two different people's bag work can

"Twist Tight Tube Corners"

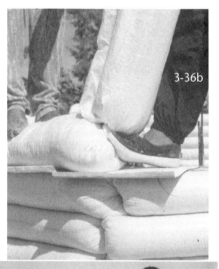

3.36a, b, c, and d: *Tubes excel at curves, but they also turn corners pretty well. Rather than ending a row of tubes at a corner, put a twist into it, turn the corner, and keep on going.* Photo credit (all 4): S.S.S.

differ over an inch (1.25 cm) in thickness if they're not on the same wavelength. The easiest way to keep the bags consistent is to have everybody gently firm the dirt in the bag with their hand after every couple of can loads. We like to fill them fairly firm. This way the wall gets taller faster. Thicker bags means fewer rows.

Closing in a Row

We divide a crew into teams and have the teams begin from either side of a window or doorway. To get a tight fit where two teams come together, fill the last two bags full enough to fill the remaining space and lower them into place at the same time (Fig. 3.37a & b). When working alone you can *hard-pack* (mild hard-assing to the full desired height) the last bag to make it extra fat, but leave it an inch (1.25 cm) shorter than the space you are filling. Drop it in. It should have enough room to fit in, but barely any space left between it and the other bags. When tamped from above, it will pound down and fill any gap.

Make sure the bags meet on equal terms. They should be flush up against each other, not one on top of the other. They should make a vertical seam where they meet, the reason being that when the row is tamped into place from above, the bags will "shoulder" into each other, limiting their movement. When they overlap (even a little), they tend to ride up onto each other and stray from the pack (Fig. 3.38).

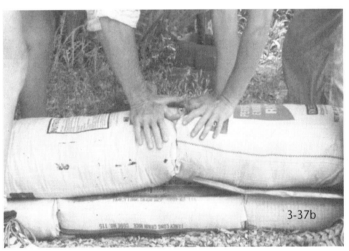

3.37a & b: *Kaki and Kay demonstrate closing in a row.*

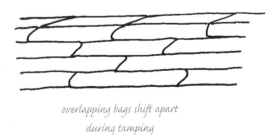

*overlapping bags shift apart
during tamping*

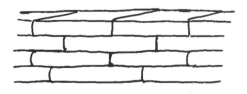

*vertical seams secure a tight fit
during tamping
FQSS!*

3.38: *Careful placement of bags will prevent these diagonal seams between bags from occurring. This overlap can cause slipping and decreased tension.*

Hard-Packing a Bag

Even when teams are filling the bags to a pretty consistent thickness, high and low spots can still show up under the scrutiny of a level. The solution is simple. Hard-pack the bags in the next row on top of the low spots and pound the puddin' out of the bags on the wall that are too high. By hard-packing, we mean pack every can dumped into the bag with your fist or a quarter-pounder. It really is just a question of over or under-filling, or pounding it harder or lighter (Fig. 3.39).

3.39 (right): *Examples of a hard-packed bag and a scooched bag.*
3.40 (below): *A pole compass with articulating arm. Note the attached level.*

Scooching

This is a technique for "soft-packing" a bag. When you want to make a really skinny bag, you may have to *scooch* it. After every two cans of dirt are added to the bag, smack the sides of the bag with both hands at the same time to compact the dirt from side to side. Be gentle when you toss in the dirt so you don't bulge the sides out again. You can also squeeze the bag stand tighter at the top, narrowing the opening. The bag will imitate whatever shape the opening resembles.

Scooched bag

Hard-packed bag

Our early bag stands were round at the top. For the 50-lb. bags it didn't make too much difference, but for the 100-lb. bags and the way-too-big bags, the giant round hole at the top caused the bags to balloon out like a laundry sack. They were insatiable beasts, swallowing entire wheelbarrows of dirt in one gulp. They were impossible to move. That's when we discovered *scooching*. Redesigning the bag stands made even the "way-too-big" bags manageable. We are living proof that humans can outwit a bag of dirt.

Construction-Size Architectural Compasses For Domes and Round Vertical Walls

The Pole Compass with Articulating Arm (Fig. 3.40)

The easiest way to maintain a precise circle during construction is by using a *compass* as a guide. For our purposes, a rigid *pole compass* works best and we use it exclusively for a variety of building designs.

The pole compass can be used for both dome building and the construction of round vertical walls like kivas, hogans, and yurts (Fig. 3.41). The pole compass uses a tall center pole with an arm attached to it that is the length of the desired radius of the building. The attached arm both rotates horizontally and pivots up and down (articulates) from its fixed point on the center pole. Most of the parts and pieces we use to build this compass are from the metal pipes and gate latches manufactured for chain link fence. These parts are available at most hardware stores in the US. Once you understand the function of these parts, however, substitutions can be made when they are not available.

A two- to three-foot (60-90 cm) long pipe is buried two feet (60 cm) into the ground. This is your base for the pole compass. Make sure it is set plumb and level. It remains in the ground and should not wobble or shift. Make sure to compact the dirt around the pipe as it is buried. Adding rocks and gravel along with the infill dirt will give it extra stability. Into this pipe is fitted a slightly smaller diameter pole that is long enough to reach the height of the second floor or loft of a dome, or the finished height of a round, plumb (vertical) wall. The fit should be snug, but loose enough that the center pole can turn in its sleeve. If the center pole binds, a few drops of oil will help it

3.41: *A 36-foot-diameter kiva-style earthbag home, going up in Wikieup, Arizona.*

turn smoothly. With a four-foot (120 cm) or longer level, check your center pole for plumb. If it is not plumb, you need to reset your base pipe. It is absolutely essential that your center pole is as plumb as possible.

Once you are happy with the plumb of this center pole, you can attach the horizontal arm. Attach the *gate frame grip* to the center pole. This "clasp fitting" allows us to adjust the height of the horizontal arm by loosening the wing nut and re-tightening it at the desired height. The "pivoting fitting" is a *fork latch* that is used as a latch for a chain link fence gate. Remove the latch piece of this assembly and replace it with a *rail end cap*. For our purpose, this part serves as holder for the horizontal arm and allows the arm to pivot up and down. Having the horizontal compass arm pivot makes moving it over the tops of door and window forms easy as it rotates around. Use a chain link *top rail* for this horizontal arm. You may have to wrap the end of the pipe with some duct tape to get a tight, non-slipping fit into the cap. You can wrap more tape around the outside of the cap and arm to maintain the compression fit (Fig. 3.42 & 3.43).

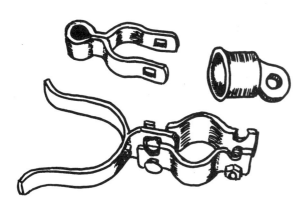

3.42: *Common chain link fittings and parts used to build a compass with articulating arm. Clockwise from upper left: Gate-frame grip; rail-end cap; fork latch.*

3.43: *Compass parts assembled onto vertical post.*

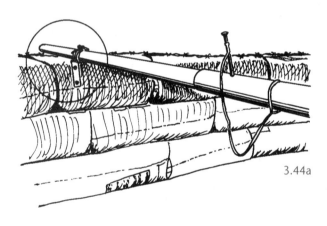

3.44a

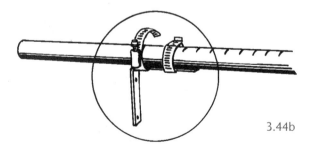

3.44b

3.44a & b: *The sliding horizontal compass arm with attached angle bracket.*

At the opposite end of your horizontal arm, attach an *angle bracket* with two hose clamps at the intended radius point of the *interior* wall. The horizontal arm itself should be longer than the fixed radius point so that it rests on top of the wall and the angle bracket barely touches the inside of the bag wall face after tamping. (Fig. 3.44a & b). We usually set it one inch (2.5 cm) inside of the determined radius to accommodate the bag expanding from tamping. After tamping the first row of bags, the finished thickness will be determined, and the compass arm will be raised on the center pole that amount for the next row. If *corbelling* the bags for a dome, the angle bracket is moved in toward the center pole the distance the next row of bags is to be stepped in. (See Chapter 12).

Onto the center of the horizontal compass arm, bind a level with duct tape. The level will show you where to tamp down the high spots and make up the difference in the low spots by overfilling the bag(s) above it on the next row. You can mark the low and high spots directly onto the culprit bags for handy reference. In most cases, placing the first row of bags on a level foundation and having all the crew fill the bags consistently with one another, makes keeping the level easy. We like to stay within one-quarter to one-half inch (0.625-1.25 cm) of level, especially if a bond beam is going on top. For domes, it is less critical but still a good idea for level window placement and maintaining the overall symmetry of the dome's shape.

A Sliding Compass Arm

A sliding insert in the horizontal arm is a useful modification for the pole compass we like to use. By placing a one size smaller diameter pipe into the horizontal arm, the arm can be made shorter or longer. When corbelling an earthbag dome, as the compass arm is adjusted upward along with the wall height, the radius is also shortening as the dome gradually closes in. Shortening this arm as the dome grows eliminates a long, overhanging pipe swinging around the work area as the dome approaches closure.

Situations may arise where parts for an articulating arm (described previously) cannot be easily obtained or fabricated. Our earliest compass arm attachments were made with rigid T-pipe connectors. Since the arm could not articulate, a sliding arm came in handy for moving the arm past protruding door and window forms. This sliding arm system works equally well for domes or vertically plumb round walls (kiva-style).

Other Compass Ideas

If a rigid pole the height of the wall or structure being built is impractical or not desired, an *expandable* sliding compass arm can be fixed at the height that the spring line begins, which is where the walls begin stepping in to create a dome.

Attach a wheel caster (with the wheel removed) to a post buried into the ground, and fasten an expanding arm to the caster (Fig. 3.45). The compass arm rotates and pivots, but since it is fixed on one end at the spring line height, it is always at an angle that changes with each row of bags. Also, as the angle increases, the length of the arm increases as well, hence the expanding compass (Fig. 3.46). A level attached to this angling arm would be ineffective, so a *water level* becomes a necessity (See Appendix for directions on how to build and use a water level).

This process becomes even more complex when using this compass to do plumb vertical walls. For vertically plumb round walls it's just easier and less time consuming to use the pole compass. The same holds true for corbelled domes, too.

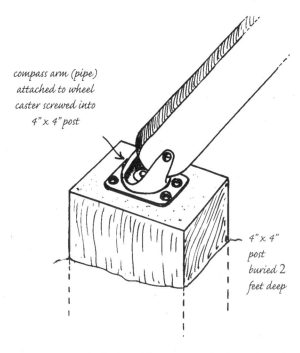

compass arm (pipe) attached to wheel caster screwed into 4" x 4" post

4" x 4" post buried 2 feet deep

3.45: *Detail for a caster wheel attachment.*

3.46: *Expandable compass fixed to center of diameter at springline determines shape by lengthening of the arm at a diagonal.*

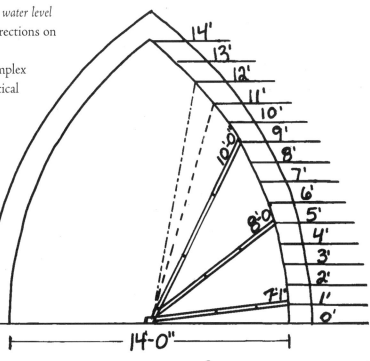

14'
13'
12'
11'
10'
9'
8'
7'
6'
5'
4'
3'
2'
1'
0'

14'-0"

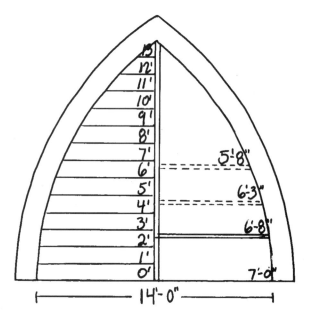

3.47: *Pole compass with adjustable length horizontal arm is used to determine radius width by shortening the length of the arm.*

For large dome projects, you can imagine that the expandable compass arm could become quite unwieldy, as the length would continue to expand to the eventual height of the dome. Thereby, a 20-foot (6 meter) wide dome would need a compass arm that expands to twenty feet (6 m) in length. A pole com-

pass arm, on the other hand, continues to shorten every row from the length of the radius. Having tried different methods ourselves, we prefer the pole compass for being simple, precise, and multi-functional (Fig. 3.47).

Finally!

Guys usually like to brainstorm about how to mechanize dirt bag construction with elaborate, rotating conveyor belts, hoppers, pneumatic tampers, and automated barbed wire dispensers. By the time you have figured out all the gizmos and the monetary investment, a good crew would have hand built an entire earthbag wall. Besides, the silence is nice. All you hear is scooping and tamping. No saws. No air guns. No banging hammers. No fumes. Just people talking, laughing, grunting, and working together as a team. People really are the most versatile equipment for earthbag construction.

As you become adept at earthbag building through planning, patience, practice, and perseverance, you too will discover new tools, tricks, and terminology that make your work better or easier or more enjoyable. And if you feel inclined to share your new-found knowledge with us, we would be honored and delighted. We are always searching for new ways to make earthbag building fun, quick, simple, and solid!

CHAPTER 4

Foundations

This has been the most exasperating chapter for us to write. Both of us had done a lot of conventional construction prior to getting involved with earthbag building. Maybe it's just us, but we both dread the idea of building a typical concrete foundation system. To us they are boring and tedious to construct. They are expensive and use up godly amounts of natural resources while pumping the atmosphere full of ungodly amounts of pollutants. Plus, they don't last very long. A typical residential concrete foundation has an average life span of 100 years. The cement eventually dissolves from efflorescence, the steel rusts, and the whole thing sucks up moisture like a sponge ... but they are sanctioned by building codes!

Our personal experience with building foundations is strongly influenced by living in a dry climate. We get sub-zero temperatures, but less than eight inches (20 cm) of annual rainfall. Since the primary foe of foundations is frost and moisture damage, our focus has been on providing excellent drainage. We are learning to adapt earthbag architecture to a moister and colder climate than our own. This chapter is designed as an informal exchange of information based on what we have experienced and what we are in the process of learning. What we offer are some examples of

4.1: *1,000-year-old Anasazi dwelling on an exposed bedrock foundation in Hovenweep National Monument.*

alternative foundation options that we can mix and match to suit our various needs and esthetic aspirations. We'll start with a brief description of a conventional foundation system and then move on to alternative adaptations.

Conventional Concrete Foundation System

Poured concrete is the most popular foundation system used in conventional construction practices. The standard procedure has been to dig a trench down to the prescribed frost line, then pour a concrete *footer* wider than the width of the wall to provide a stable base. Within this footer, steel rebar is suspended to provide tensile strength, as concrete alone is brittle. On top of the footer, a *stem wall* is poured (with additional rebar for reinforcement) equal to the width of the finished wall of the structure and tall enough above grade to keep the wall dry (Fig. 4.2).

Frost depths vary with the climate from non-existent to permanent. The average depth that frost enters the ground here in Moab, Utah is 20 inches (50 cm). The average frost depth in Vermont is four feet (120

A FOUNDATION PERFORMS SEVERAL FUNCTIONS:

- A solid footing to distribute the perimeter weight of the structure evenly over the surface of the ground.
- A stable base that defies the up-heaving and settling forces caused by freeze/thaw cycles in cold, wet climates.
- A protective perimeter that guards the lower portion of the walls from erosion and moisture damage.
- A means to anchor a structure to the ground in response to severe weather conditions such as high winds, flooding, earthquakes, etc.

cm). The logic is that by beginning the foundation below the frost depth, the foundation rests on a stable base, free from the forces of expansion and contraction occurring from freeze/ thaw cycles commonly called *frost heaving* Moisture freezing below an insufficient foundation depth can result in upheaval of the structure causing cracking of the walls or annoying things like forcing all the doors and windows out of alignment. Conventional construction logic believes that deep frost lines require deep foundations.

This logic poses a challenge to most alternative architecture, as new energy priorities have increased the width of the walls to as much as two to three feet (60-90 cm) thick. To adapt conventional poured concrete foundations to thick earthen walls would defeat the resource and cost effectiveness of building them in the first place. So it makes sense to innovate a foundation system suited to thicker walls.

One suitable foundation system is inspired by a 1950's design by Frank Lloyd Wright, devised as a way to lower construction costs for foundations built in

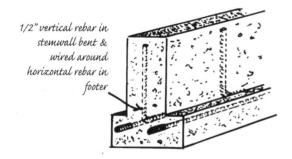

1/2" vertical rebar in stemwall bent & wired around horizontal rebar in footer

4.2: *Cross-section view of a typical concrete foundation and stem wall.*

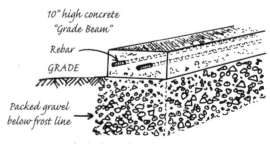

10" high concrete "Grade Beam"

Rebar

GRADE

Packed gravel below frost line

Frank Lloyd Wrights "Floating Footer Foundation"

4.3: *Floating footer on a rubble trench foundation.*

deep frost climates. It is called the *rubble trench* foundation with a *floating footer*. Instead of filling the frost depth of the foundation trench with poured concrete, he used rubble stone and packed gravel as the supportive base and limited the concrete work to a ten-inch (25 cm) high concrete, steel reinforced *grade beam* that rests on the surface of the rubble trench. To avoid frost heaving, the rubble stone allows drainage below the frost line where water can continue to percolate. This combination rubble trench/grade beam acts as the footer and the stem wall. Expenses and resources are reduced with the substitution of packed rock for an entire poured concrete foundation (Fig. 4.3).

Another successful foundation system being utilized by HUD (Housing and Urban Development) is called a *shallow, frost-protected* foundation. Since one of the main objectives of a foundation is to defy freeze/thaw action, another solution is to inhibit frost penetration from entering the foundation with buried exterior insulation. An insulated foundation also helps reduce heating costs, as an average of 17 percent of a home's warmth can escape through the foundation (Fig. 4.4). You can visit HUD's website for more detailed information and to find out the specs on using an insulated frost-protected foundation system for your climate. Refer to the Resource Guide for the website location of HUD.

Of course, conventional concrete foundations and stem walls work with earthbag walls as well.

Earthbag Rubble Trench Foundation Systems

For the sake of simplicity and function, we use a streamlined version of the rubble trench foundation system for building freestanding garden walls, and a simplified, low-tech version of Frank Lloyd Wright's floating footer system for a house. Either system can be appropriate for an earthbag dwelling depending on the climate. The main difference between a conventional poured concrete foundation and the foundation systems we have adopted for earthbag walls is that our systems are built of individual units rather than a continuous beam.

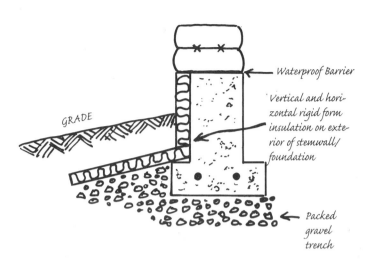

4.4: *Shallow, frost-protected foundation.*

A continuous concrete foundation is a fairly recent invention. Humans have been building "alternative" foundations for 10,000 years. When we observe history, the oldest surviving structures throughout the world are sitting on individual stone blocks and packed sand and gravel. Some are "cemented" together with a mud mortar like the 800 - 1,200 year old Anasazi ruins found in the Four-Corners region of the Southwestern United States. The Romans and Greeks built whole empires with rubble rock held together with the glue of lime. In the Northeastern United States, 200-year-old Vermont timber frame barns are still sitting on dry-stack stone foundations. Neither the 600-year-old stone Trulli villages in Italy, nor the 300-year-old cob cottages off the coast of Wales have a lick of steel or cement in them.

The other difference is that we use gravity as our anchor to the foundation/stem wall rather than bolts or impaling the bags with rebar (a common choice for attaching strawbale walls to a concrete stem wall). Instead, we rely on the massive weight and strategic design of the walls to keep the building stable. (Refer to Chapter 5).

Our Basic Rubble Trench

For garden and privacy walls in our dry climate, we dig a trench about four to six inches (10-15 cm) wider than the width of the proposed finished wall, and

4.5a

4.5b

about 12 inches (30 cm) deep (about one-half of our average frost depth). In wetter, colder climates it may be necessary to dig below the frost line. The trench is filled with coarse rock progressing to smaller gravel toward the top of the trench. Any sand in the mix should be clean and coarse (avoid silty and clay-rich soils). Whatever we can get that compacts well and still provides good drainage will work. Spray with water during installation to help the gravel and any sands to compact better. This is the basis of the rubble trench foundation (Fig. 4.5a & b).

Concrete Earthbag Stem Wall

Our idea of a concrete stem wall for an earthbag wall is to fill the first two to three rows of bags on top of a rubble trench with concrete. Marty Grupp built a "fast-food mentality" concrete earthbag stem wall for two serpentine walls in front of a small apartment complex. He laid a row of Quick-crete bags, perforated each bag and soaked them with water. He laid down two strands of barbed wire and another row of Quick-crete bags, perforated and sprayed them with water and called it done. He figured that by the time he ordered all the materials and the mixer and the extra help to pour all the concrete, he could lay prepackaged bags of Quick-crete himself for just about the same amount of money; and he did! (Fig. 4.6).

Our biggest challenge has been coming up with an entirely cement-free foundation/stem wall. As far as architectural time scales go, "portland cement" is a fairly recent development. In 1824, Joseph Aspdin, an English bricklayer, patented a process for making what he called portland cement, with properties superior to earlier varieties. This is the cement used in most modern construction. The use of cement has greatly increased in modern life. It's used for foundations, walls, plasters, blocks, floors, roofs, high rises, bridges, freeway overpasses, highways, sidewalks, swimming pools, canals, locks, piers, boat slips, runways, underground tunnels, and dams ... to name a few.

4.5a, 4.5b: A simple rubble trench showing (A) a rubble rock base, (B) topped off with clean, well-tamped gravel.

It takes a lot of produced power (embodied energy) to produce cement. According to *The Adobe Story* by Paul G. McHenry, in the US it takes "four gallons of gasoline or diesel fuel to produce one bag of cement, while contributing over 8 percent of the total carbon dioxide released into the atmosphere." Worldwide, cement production accounts for 12 percent of that total, or one ton of carbon dioxide for every ton of cement produced. It is in our best interest as a species to learn to minimize our dependence on cement. With these cheery thoughts in mind, on our way to cement free foundations, let's look at some reduced-use, cement stabilized options.

Stabilized Earth Stem Walls

An alternative to using full strength concrete is to fill the first two or three rows of bags (the stem wall bags) with a *stabilized earth* mix. Stabilized earth is a method of making a soil resistant to the effects of moisture by adding a percentage of a stabilizing agent. As Joe Tibbets states in his excellent reference, *The Earthbuilders' Encyclopedia*, "The advantage of using a stabilized earth … is that we can use the soil we already have available for our walls instead of importing washed concrete sand and gravel. As is the case of cement, it uses a lower percentage than for full strength concrete reducing cost of materials." Common stabilizing agents are cement, asphalt emulsion, and lime. Generally most soils suitable for stabilization are coarse, sandy soils. There are exceptions and all three stabilizing agents function in different ways.

The following information on cement and asphalt stabilization is adapted from Joe Tibbet's *Earthbuilders' Encyclopedia*.

Cement acts as a binder, literally gluing the particles together. Cement provides adhesion as well as additional compression strength. By adding a predetermined amount of cement (anywhere from 6 percent-15 percent depending on the particular soil), cement fully-stabilized earth (aka: soil cement) can be used as a way to minimize the use of cement while making use of cement's ability to remain stable when it comes in contact with water. Cement is more effective

4.6: *Marty Grupp's Quick-crete stem wall.*

4.7: *Because of the lack of tensile strength (e.g. rebar), a continuous tube is more likely to crack in a cold, wet climate, so using tubes would probably work best in a dry or frost-free locale.*

with soils low in clay content with a coarse, sandy character. Cement mixed with soils high in expansive clay is less effective as the two tend to oppose each other, thereby compromising their bonding strength. Using tubes filled with cement stabilized earth for foundation/ stem walls creates an interesting effect because there are fewer seams (Fig. 4.7). Once cured, the fabric can be removed to reveal a sculpted stone appearance. This looks particularly exotic in a serpentine wall. The exposed soil cement can be stained or lime-washed any color desired.

Two advantages of using a soil-cement are: reduced ratio of cement to aggregate and (if the soil is suitable) cost savings from not having to purchase more expensive washed gravel and sand needed for standard concrete.

Asphalt emulsion provides a physical barrier to the passage of water. A properly prepared asphalt stabilized earth mix will not absorb more than 2.5 percent of its own weight in water. It functions by surrounding the particles of clay-bound sand clusters with a water-resistant film. The percentage of asphalt emulsion added to soil for stabilization ranges from about 3 percent to a maximum of 6 percent. Any more than that jeopardizes the integrity of the soil. Because it comes in a liquid form, asphalt emulsion makes the mix wetter and may require more time to set up before continuing the earthbag wall system. This is our least favorite form of stabilization. Other than some early experimentation, we do not use it. Asphalt emulsion is a carbon-based fossil fuel by-product, and a known carcinogen.

Lime Stabilization. Of the three common stabilizers, lime is the one most compatible with clay-rich soils. When we speak of lime, we are referring to *building lime* and not *agricultural lime*. The most familiar form of building lime in North America is Type S - Hydrated Lime that comes in a dry powdered form in 50-lb. (22.2 kg) bags. It is most commonly used as an additive to cement stucco and cement-based mortars to enhance workability and inhibit moisture migration.

Lime is made by firing limestone to produce calcium oxide by burning off the carbon. This calcium oxide (also referred to as *quicklime*) is then reacted with specific amounts of water to produce building lime. (Agricultural lime is simply powdered limestone in its natural, unfired state.) The complexities of lime are fascinating and worth taking the time to research and learn about. (See the Resource Guide for suggested reading about lime).

For the purpose of soil stabilization we will focus on the use of Type-S lime hydrate available in the US at most lumber yards, building supply warehouses, and wherever cement products are sold. One critical factor worth mentioning is the need to acquire lime in as fresh a state as possible as it weakens with exposure to moisture from the air over time. Wrapped in plastic, fresh off the pallet from the lumber yard, it ought to still have some life to it. Purchasing lime directly from the manufacturer and sealing the bags in plastic garbage bags until needed helps ensure the lime's potency.

Interestingly enough, lime used as a soil stabilizer for road work was pioneered in the US in the 1920's. Thousands of miles of roads have been constructed on top of lime stabilized soils. As noted by Hugo Houben and Hubert Guillaud in their book, *Earth Construction,* "the Dallas-Ft. Worth airport was constructed over 70 square kilometers using lime as a base soil stabilizer."

Lime interests us as a soil stabilizer for several reasons. Ton per ton, it takes one-third the embodied energy to produce lime than it does cement. During lime's curing process, it reabsorbs the carbon dioxide it gave off when it was fired. In a way, in cleans up after itself. Cement, on the other hand, pumps a ton of carbon dioxide per ton of cement produced into the atmosphere and leaves it there. Lime is the lower impact choice for stabilizing a soil.

Here is a simplified explanation of how lime stabilization works. Lime reacts with clay in two significant ways. First, it agglomerates the fine clay particles into coarse, friable particles (silt and sand-sized) through a process called *Base Exchange*. Next, it reacts chemically with available silica and aluminum in the raw soil to produce a hardening action that literally glues all the particles together. This alchemical process is known as a *pozzolanic reaction*. Other additives that cause this chemical reaction with lime are also called *pozzolans*. The origin of the term pozzolan comes from the early discovery of a volcanic ash mined near Pozzolano, Italy that was used as a catalyst with lime to produce Roman concrete. Venice,

Italy is still held together by the glue of lime reacted with a pozzolan. In addition to volcanic sands and ash, other pozzolans include pumice, scoria, low-fired brick fines, rice hull ash, etc. Any of these can be added to fortify a lime stabilized earth.

Lime reacts best with montmorillonite clay soils. For stabilizing stem walls, the optimum soil is strongest with a fair to high clay content of 10 percent-30 percent and the balance made up of well-graded sands and gravel to provide compressive strength. Often the material available as "road base" or reject sand at gravel yards is suitable for lime stabilization.

Another distinctive advantage of lime is that lime stabilized soil forms a water resistant barrier by inhibiting penetration of water from above (rain) as well as capillary moisture from below. This indicates that lime stabilized earth is less likely to require a significant *capillary break* built in between a raw earth wall and a lime stabilized stem wall (more about capillary break later in this chapter).

Experimentation will determine whether the soil is compatible with lime and what appropriate ratio of lime hydrate to soil will be needed to achieve optimum results. Every dirt will have its favorite ratio of lime to soil. In general, full stabilization occurs with the addition of anywhere from 10 percent-20 percent (dry volume) lime hydrate to dry soil, depending on the soil type. Adding 5 percent-25 percent of a pozzolan (determined by tests) provides superior compression strength and water resistance.

To facilitate proper curing, lime stabilized earth must be kept moist over a period of at least two weeks — three is better. The longer it is kept moist the stronger it sets. For use in earthbag stem walls, this is easy to achieve by covering the rows with a plastic tarp and spritzing them occasionally with water. Double-bagging the stem wall bags will also help to retain moisture longer. *The moisture curing period is essential for creating the environment necessary to foster the pozzolanic alchemy that will result in a fully stabilized soil.* A fully stabilized soil is unaffected by water and will remain stable even when fully immersed. It's worth the effort to achieve especially if you are building in a wet climate.

Mixing procedure for cement or lime stabilized soil.

Pre-mix the soil and cement or lime in a dry or semi-damp state. Mixing them dry achieves complete mixing of the two materials. Mixing can be done by hand in a wheelbarrow or in a powered cement or mortar mixer. Once all of the dry ingredients are thoroughly integrated, water can be added. A slightly moister mix than that of a typical rammed earth mix is needed. Add enough water to achieve about 20 percent moisture, or enough water that the moisture will slightly "weep" through the weave of the earthbags when tamped. Keep damp for as long as possible to cure properly.

Moisture Barriers, Vapor Barriers, and Capillary Breaks

To avoid any confusion, we want to explain the difference between *moisture barriers* and *vapor barriers* before we get into how they are used. A *moisture barrier* inhibits moisture penetration (water in the liquid state), but allows vapor (water in the gaseous state) to transpire. Moisture barriers are generally used in an external application, such as "Tyvek wrap." A *vapor barrier* impedes water migration in both the liquid and gaseous states. A vapor barrier is also referred to as a *waterproof membrane*. A good example of a vapor barrier is plastic sheeting.

Although cement will remain stable when in contact with water, it has a notorious habit of wicking moisture up from the ground resulting in water migration into the earthen wall it is designed to protect. Many earthen walls have succumbed to failure due to well-meaning yet incompatible restoration repair jobs using cement to stabilize historic adobe missions throughout the Southwestern United States. Earthen walls retain their integrity as long as they stay dry or can dry quickly when they do get wet.

Cement wicks moisture as well as inhibits evaporation. Water goes in but is slow to come out. Water likes to travel and will search for an outlet even if it means defying gravity by migrating into the more porous, raw earthen wall above. Water's rising by

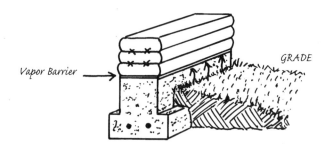

4.8: *Capillary action: Without a capillary break, moisture in the ground is wicked up by the concrete and absorbed into the more porous earthen wall above.*

4.9: *The waterproof membrane prevents capillary action, inhibiting water absorption by imposing a barrier.*

absorption into a more porous substrate above is described as *capillary action*, much like how a sponge soaks up water.

Conventional wood frame construction is required by code to install a vapor barrier in between the top of the concrete stem wall and the wooden sill plate that the stud wall framework is attached to. It can be a roll of one-eighth-inch (0.3 cm) closed cell foam, heavy gauge plastic sheeting, tarpaper, or a nontoxic alternative liquid sealer, like DynoSeal made by AFM products, slathered on top of the surface of the concrete stem wall (Fig. 4.8 & 4.9).

However, a waterproof membrane (vapor barrier) designed to inhibit water absorption from below can also prevent drainage from above. If for some reason water was to enter the wall from above the foundation (from leaky roofs, windows, or plumbing) it could

dribble down onto the surface of the vapor barrier, pool up and be wicked into the wall it was designed to protect. There is a lot of debate about the use of vapor barriers and cement in general as both materials have shown evidence of retaining or diverting dampness to organic building materials, causing moisture problems.

An alternative approach is to design a *capillary break* that prevents moisture rising from below as well as providing drainage from potential moisture invasion from above. This can be achieved by creating large enough air spaces so that water is unable to be absorbed. Double bagging the woven poly bags and filling them with three-quarter-inch (1.9 cm) gravel can make a simple capillary break. Although we feel gravel-filled bags would make an effective capillary break, we have yet to experience how they would hold up over time. Since the gravel is held in place by relying entirely on the bag, in addition to doubling the bags, make sure to keep them well protected from sunlight immediately after installation to ensure the full benefit of their integrity.

Another option for a capillary break is a couple of layers of flat stones arranged on top of the concrete or stabilized stem wall in a way that allows air passage to occur between the rocks. Rock should be of an impermeable nature (rather than a porous type like soft sandstone) that will inhibit moisture migration (Fig. 14.10a & b). The entire stem wall can be built out of stone, dry-stacked directly on top of a rubble trench. A rubble trench is, in itself, a type of capillary break.

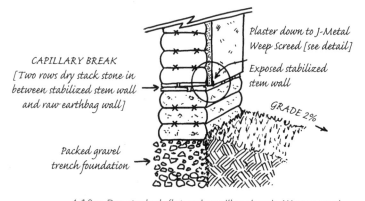

4.10a: *Dry-stacked, flat rock capillary break. Weep screed between earth plaster and stabilized stem wall.*

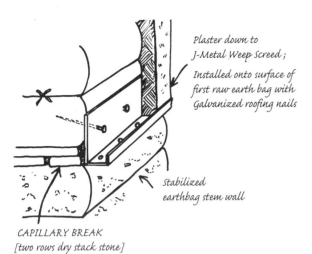

Plaster down to
J-Metal Weep Screed;

Installed onto surface of
first raw earth bag with
Galvanized roofing nails

Stabilized
earthbag stem wall

CAPILLARY BREAK
[two rows dry stack stone]

4.10b: J-Metal weep screed detail.

4.11 (above): *Plastering down to ground level risks wicking moisture into the walls, causing spalling and other moisture-related damage.*

4.12 (below): *Traditional and alternative foundation/stem walls.*

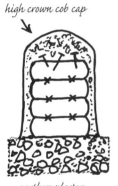

high crown cob cap

earthen plaster
on gravel trench

stone gable cap

sloping stone veneer

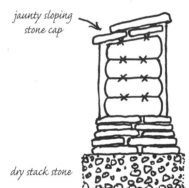

jaunty sloping
stone cap

dry stack stone

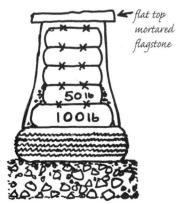

flat top
mortared
flagstone

50 lb

100 lb

good year tire filled with gravel
on rubble rock

stabalized earth, rebar anchors,
two strands
barbed wire

sloping
rock drip
edge

"urbanite" recycled busted concrete
dry stacked or mortared

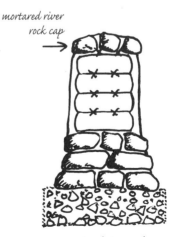

mortared river
rock cap

mortared river rock

Weep Screed (often referred to as "J-metal" in the building trades) is a platform for the wall plaster to come down to. It is a mini-capillary break that protects the plaster from wicking moisture up from the ground or a concrete or stabilized foundation/stem wall (Fig. 4.11).

An alternative to "J-metal" is five-eighths-inch (1.5 cm) or three-quarter-inch (1.9 cm) hose secured in between the stem wall bags and the raw earth bags with either tie wires or long finish nails. We have used both soaker hose and black "poly-tubing" (used commonly for irrigation) as a flexible weep screed for curved walls. J-metal, however, can be successfully "clipped" to conform to curved walls as well.

The reason we mention something intimately related to plaster at this time is to point out the fact that all things are tied to and dependent on each other. The big advantage to prior planning is that it allows you to address a situation early on in the construction process to make a later task easier. Think ahead!

Traditional and Alternative Foundation/Stem Walls

The stem wall is the most vulnerable part of the foundation system since it is the most exposed to the elements. This is the area where splash occurs and wet leaves cluster, that grass migrates toward, and where microorganisms in the soil try to munch the wall back into compost. In a really dry climate (less than ten inches [25 cm] of annual rainfall), we can get away with placing the raw, natural (non-stabilized) earthbags directly on top of the rubble trench with a yearly maintenance of earth plaster. Adding a protective rock veneer as a splashguard on the exterior of the natural earthbags will increase their durability. Other options include dry-stack stone, mortared stone with earthen or lime base mortar, stabilized adobes, fired brick, recycled broken concrete slabs, and gravel-filled or rammed earth tires (Fig. 4.12).

A *tire stem wall?* Mike Reynolds, innovator and designer of the "Recycled Radial Ranchos," or Earthships as they're better known, refers to discarded tires as "indigenous." Old tires can be found just about everywhere, so we might as well make use of them. At the 1999 Colorado Natural Building Workshop in Rico, Colorado, Keith Lindauer (an avid Earthship builder) prepared an impressive terraced rammed tire foundation for us to build an earthbag garden wall onto. For the most part, rammed earth or gravel filled tires are a great way to turn an indigestible man-made artifact into a durable stem wall. Keep in mind that many more options exist for innovating new uses for old materials (Fig. 4.13).

Insulated Earthbag Foundation/Stem Walls

The colder and wetter the climate is, the more a foundation will benefit from the addition of exterior insulation. Insulation is most effective when placed on the exterior of a foundation, as it provides a warm air buffer in between the earthen mass and the outside temperatures. In a cold climate, insulation increases the efficiency of an earthen wall's mass allowing it to retain heat longer and reradiate it back into the living space.

The type of rigid foam we prefer to use is the high density white bead board made from expanded polystyrene (EPS). As of this writing, it is the only rigid foam made entirely without chlorofluorocarbons

4.13: *Tire foundation/ stem wall used to support an earthbag wall at Colorado Natural Building Workshop in Rico, Colorado.*

(CFC's) or hydro-chlorofluo-rocarbons (HCFC's). It comes in two densities with the higher density the better choice for below ground applications. It has environmental drawbacks, as substantial energy is used to produce it, but can be recycled to a certain extent. High density bead board has a R-value of 4.35 per inch (2.5 cm) of thickness. We reserve its use for perimeter insulation around a foundation (Fig. 4.14). We'll talk later about a more natural (but less available) technique for increasing the insulation of an earthbag. For now, let's look at the advantage of rigid foam insulation.

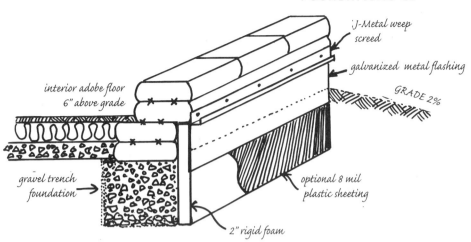

4.14: *Insulated earthbag foundation.*

One advantage to using rigid foam as perimeter insulation, is that by protecting the earthbags from moisture, we can use a raw rammed earth mix and so avoid using any cement in the bags. For below ground applications, high density rigid foam has a fairly high compression strength that is able to resist the lateral loads imposed on it by backfill. The exposed foam insulated stem wall can be sealed with heavy (8–10 mil) plastic sheeting followed by metal flashing, rock facing, bricks, or packed, sloping gravel to protect the foam from UV deterioration.

For round walls, we apply two layers of one-inch (2.5 cm) rigid foam because it is flexible enough to bend around a curve. Be sure to alternate the seams so there is no direct path for water to migrate (Fig. 4.15). If you are uncertain of the water resistance of your rigid foam, or you want greater water protection, there are several commercially manufactured water resistant membranes available. Heavy gauge polyethylene sheeting or butyl rubber will help inhibit the transport of moisture into your insulated foundation from groundwater, storm runoff, or spring thaw. This is very smart protection for bermed or buried structures where moisture is prominent. Lumberyards, farm co-ops, and catalogues are great sources for heavy 8–20 mil agricultural grade polyethylene sheeting.

For extreme external moisture protection, *pond liner* material, EPDM (ethylene propylene diene monomers), roof sealer, or a heat-sealed bitumen fabric can be wrapped around the exterior insulation and then back-filled into place. Pond liner is a heavy gauge black butyl rubber material usually reinforced with a woven grid imbedded within it to resist tearing. It has a 50-year (average) lifespan. Protected from the sun, it would probably last much longer.

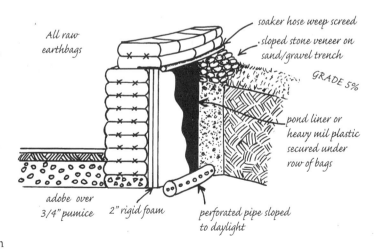

4.15: *Curved, insulated foundation with interior sunken floor.*

Pumice/Earthbag Insulated Foundation

Pumice is a light, porous volcanic rock often used for scouring, smoothing, and polishing. It is rock filled with tiny air pockets. As a result, it works well as an insulative layer. By premixing pumice rock with suitable rammed earth quality dirt at a 50:50 proportion, we have made earthbag blocks that weigh one-third their original weight. We haven't done any "official" tests on the insulative quality of these mixes. We only assume that with the additional trapped air spaces we are getting some kind of insulating effect (Fig. 4.16).

4.16: *The pumice/earthbag cures hard and flat and holds together like a typical rammed-earth mix.*

By combining the pumice with 50 percent earth, we are still able to tamp the mix into a compacted block that holds together like rammed earth. Filling the bags with pumice alone produces a lumpy bag full of loose material that refuses to compact while lacking the weight that we rely on for gravity to hold it in place. We prefer to maintain the structural integrity of the wall system first, and then figure out ways to address insulating options.

The pumice should be of a size range between three-quarter-inch (1.9 cm) up to one-inch (2.5 cm) in diameter. According to Tom Watson, the designer of the "Watson Wick" (a natural gray and black water living filtration system), a small-sized pumice will wick moisture up from the ground like a sponge, whereas a larger diameter pumice will drain moisture away. *Scoria*, another type of rock produced from volcanism, may be substituted for pumice. Experiment with the ratio of earth to scoria to find what works best for your project.

Large quantities of pumice and good clean tampable earth can be premixed with a backhoe or tractor loader and heaped into a pile to be moistened, tarped, and ready for wall building as with a regular earthen soil. Extra water will need to be added to account for the increased amount absorbed by the pumice. For stem walls, or any place where moisture may be a problem, the pumice/earth combo can be stabilized with the same percentage of cement or lime suitable for the soil being used.

Design Considerations for Bermed and Buried Structures

For clarification, when we refer to a "buried structure," we tend to think more in terms of a sunken floor, at most about four feet (120 cm) deep. A "bermed structure" is usually buried into a slope (preferably with a southern exposure). A bermed structure can also be built at grade level on a flat plain with its north side buried by piling up earth around it, making a sort of man-made slope. A structure that is completely underground we refer to as "subterranean."

Round Is Sound

As far as using earthbags as an alternative foundation system to conventional concrete, we have one very strong recommendation to make. Build round when you build underground. Bermed and buried walls undergo tremendous stress from the surrounding earth as the walls are backfilled as well as over time as the world settles in around them. When the earth exerts pressure onto the walls of a round structure, the compression is equally distributed throughout the full circle (Fig. 4.17).

The same principle is used to contain water behind a dam. Most dams have a curve in them that backs into the water. The water puts pressure on the dam and the shape of the dam distributes the pressure along its whole surface. A linear wall, on the other hand, may be strong at the corners, but is weak along the straight runs. Under pressure, it is far more likely to "blow out," or rather, collapse inward over time.

The earthbag wall system is designed to work in conjunction with the forces of compression to maintain its structural integrity. Adding a curve in the wall is the simplest way to achieve this. A square is fair, but a curve has nerve (Fig. 4.18).

As we have demonstrated in the illustrated step-by-step wall system guide, straight walls need to be buttressed and have all of their exposed end bags *hard-assed*. A curved wall is not only stronger; it entails less time and energy to build. A linear design compromises structural integrity, uses more time, energy, and resources to construct, but allows the sofa to fit up against the wall. Pick your own priority.

Site Evaluation

Evaluating the building site for an earthbag structure follows all the same criteria as any other structure. If you are on a flat plain with decent drainage or a southern slope, you have an ideal opportunity to partially bury or berm your structure. Fairly stable soils and sandy soils are ideal for buried structures.

4.17: *The Anasazi built round, buried kivas that still remain intact after 1,000 years, while a white man builds a square root cellar that falters before 100 years have passed. Does this mean a circle is ten times stronger than a square?*

4.18: *Penny Pennel's 36'- Diameter Bermed Earthbag Kiva Southern, AZ.*

4.19a, b, and c:
French drain installation details.

porous filter fabric

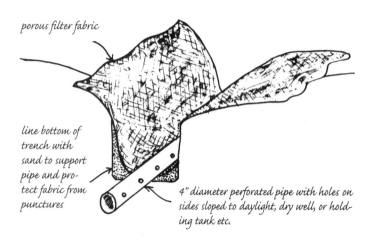

line bottom of
trench with
sand to support
pipe and pro-
tect fabric from
punctures

4" diameter perforated pipe with holes on
sides sloped to daylight, dry well, or holding tank etc.

4.19 a

fold filter fabric over top of cobbles

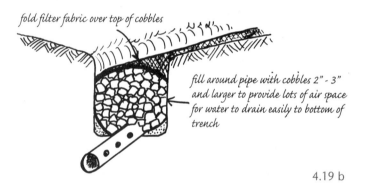

fill around pipe with cobbles 2" - 3"
and larger to provide lots of air space
for water to drain easily to bottom of
trench

4.19 b

pile more cobbles and rocks on top of
exposed filter fabric

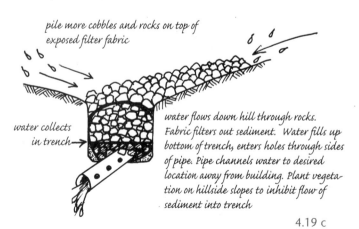

water collects
in trench

water flows down hill through rocks.
Fabric filters out sediment. Water fills up
bottom of trench, enters holes through sides
of pipe. Pipe channels water to desired
location away from building. Plant vegetation on hillside slopes to inhibit flow of
sediment into trench

4.19 c

When planning to "dig in," bury, or berm an earthbag structure, avoid sites located in an area with a high water table, flood plains, natural drainage areas ("dry" washes and intermittent streams), bogs, swamps, highly expansive clay soils, and steep slopes prone to mudslides and rock falls. Even with all these careful considerations, a bermed structure will require additional earthwork around the perimeter of the structure to ensure proper drainage. The installation of a *French drain* or *swales* will divert water around and away from the structure. Extra care must be taken to ensure adequate drainage exists around the entire perimeter of a buried earthbag building (Fig. 4.19a, b& c). (For more information about French drains and swales, consult the Resource Guide at the back of this book).

Earthquake Resilient Foundations

History has shown us that a foundation can be constructed of individual stacked units like rock and brick just as successfully as a poured concrete foundation. A poured concrete foundation relies on rigidity as a means to provide stability in regard to earthquakes. We can try to overcome nature through resistance or we can go with the flow and flex right along with her. The latter has been the design preference of choice for thousands of years throughout the Middle East and Asia in the most active earthquake regions of the world.

As far as earthbag building is concerned, Nader Khalili, innovator of the earthbag method, had tests conducted on his workshop structures to simulate earthquake movement. The tests were done in accordance with ICBO standards for a structure in earthquake Zone 4. This is recognized as the highest earthquake zone in the United States. Test results far exceeded the limits set by the ICBO, and in fact, the testing apparatus began to fail before any deflection was observed in the buildings tested.

Tests conducted at the University of Kassel in Germany conclusively prove that in comparative studies of square and round rammed earth structures, round structures show much higher stability in earthquake impact tests. The report went on to state that "it is advantageous if the resonant frequency of the house does not match the frequency of the earth movement during an earthquake." This implies that heavy houses built with solid construction (as in the case of earthbags) should not be attached to a rigid foundation. Light houses (such as frame construction) perform better attached to a solid foundation. In an earthquake, buildings are mainly affected by the horizontal acceleration created by the movement of the earth. A massive structure, independent of the foundation is able to move independently of the foundation. Therefore, in an earthquake, the ground stresses transmitted to the foundation are not carried through to the building. A light-weight structure relies on its elasticity to counter the seismic forces on the foundation it is attached to.

If you intend to build an earthbag building in a high seismic zone, intensive research should be undertaken, and a full knowledge (or at least a structural engineer's blessing) of what is required should be employed. It's probably a good idea to use a round design, too. Designing buildings for earthquake resistance is beyond the scope of this book, but as more research is done in this area, indications are that earthbag buildings could prove to be a low-cost, low-impact alternative to present day conventional seismic construction practices.

CHAPTER 5

Structural Design Features for Earthbag Walls

Every construction medium incorporates specific design principles to get the optimal performance from the material being used. Timber, stud frame, and post and beam incorporate diagonal bracing and crossties to provide shear strength. The dimensions of the lumber and spacing dictate load-bearing capabilities, etc. No building material is immune to nature's governing principles. Even rock is affected by frost heave.

Earthbag building is still in its infancy, so it remains open for exploration. The design principles that we have incorporated into earthbag building are simple, common sense strategies inspired by FQSS, observation of successful indigenous building techniques, and some of the current provisions to adobe building codes, especially those of the state of New Mexico where they have a long-standing tradition of building with earthen materials.

A list of the fundamental structural principles for building vertically plumb earthbag walls is as follows (note that domes are in a category all to themselves). (Refer to Chapter 11).

5.1: *Linear, freestanding walls require lateral support (buttressing) every 12 feet on center.*

5.2: *Various shapes that provide lateral stability —*
buttressing for linear walls or curves.

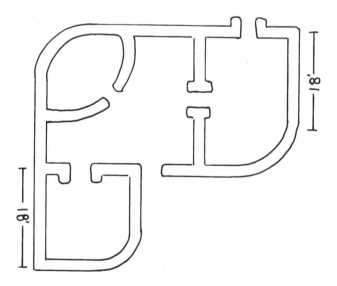

5.3: *Buttresses, corners, intersecting earthbag or stud-*
frame wall, and sufficient curves are all
considered effective forms of lateral support
in a structure.

Height Limitations

We suggest building single-story earthbag buildings no more than ten feet (3 m) in height. This is based solely on our personal experience of building 15-inch (37.5 cm) wide earthbag walls ten feet (3 m) tall. If you would like to continue building a second story using earthbags, a concrete bond beam would be advisable, placed between these two levels. An alternative to a concrete bond beam is to double the thickness of the first story walls, or use larger width bags, or built-in lateral support (buttressing).

Lateral Support (Fig. 5.2)

Lateral support is a method of bracing a straight run of wall to give it extra stability. A linear wall is fairly easy to push over as it offers very little resistance to pressure exerted from one side or the other. Adding a wall that intersects a linear wall is a type of lateral support.

Adobe codes typically require buttressing (lateral support) every 12 feet (3.6 m) on center for freestanding walls, and 24 feet (7.2 m) on center for walls intended as structures. We follow the adobe code criteria for freestanding walls. However, since earthbag walls are built "green," they tend to have some flex to them during construction until the earth cures. As a consequence, we feel more comfortable including buttressing for a structure every 18 feet (5.4 m), depending on the thickness of the wall and the soil type (Fig. 5.3).

Interior walls benefit from buttresses as well. They can be split onto either side of a dividing wall to reduce their profile. They add esthetic interest as built-in nooks or support for shelving. Curved walls perform the task that a buttress does with more grace and efficiency (Fig. 5.4).

Height to Width Ratio

In addition to lateral support, buttresses (as well as curves) follow criteria for determining their *height to width ratio*. For every foot (30 cm) of wall height, add six inches (15 cm) of width to the wall, either as total thickness, or as a curve or buttress. In other words, the

5.4: *Sufficiently curved walls are laterally self-supporting.*

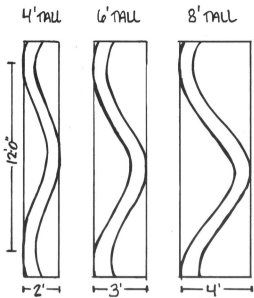

5.6: *Examples of height to width ratios for curved, freestanding walls. (Bird's eye view).*

height to width ratio is two to one expressed as 2:1, height to width (Fig. 5.5).

Curved walls follow the same criteria by squiggling in a curved pattern that fulfills the height to width ratio within every 12 feet (3.6 m) of length (Fig. 5.6). Another example is that you can also increase the total thickness of the wall by making a six foot (1.8 m) high wall that is three feet (0.9 m) thick, or a four-foot (1.2 m) wall two feet (0.6 m) thick, etc. Buttressing was obviously devised to reduce materials and labor.

Round Walls

Round is sound. A round structure is the most stable of wall shapes and is particularly well suited for below ground applications and in earthquake and hurricane prone environments. Build round when underground, or at least add a curve into a back bermed wall (Fig. 5.7).

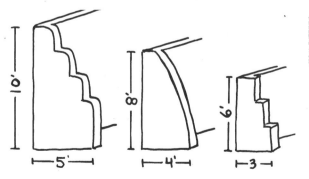

5.5: *A ten-foot (3m) tall wall will need a buttress with a total depth of five feet (1.5 m), spaced at every 12 feet on center. An eight-foot (2.4 m) wall will need a buttress four feet (1.2 m) deep, and a six foot (1.8m) tall wall requires a buttress that is three feet (0.9 m) deep.*

MARLENE WOLF

5.7: *The ultimate form of buttressing is a complete circle.*

Proper Placement of Barbed Wire

Make sure to incorporate the barbed wire into all the buttresses and wrap it around any corners ending it at the box form openings (rather than at a corner). If you are building earthbag walls that are two bags wide, lay the barbed wire in a repeating figure eight pattern to help link the side-by-side bags together. Remember to stagger the seams where bags meet (Fig. 5.8 & 5.9).

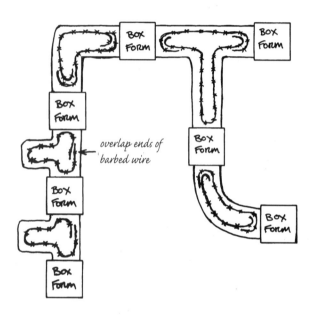

5.8: *To maintain as much tensile strength as possible, continue barbed wire around corners, and integrate into buttresses and intersecting walls.*

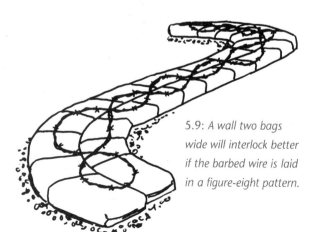

5.9: *A wall two bags wide will interlock better if the barbed wire is laid in a figure-eight pattern.*

Interlock Corners

When building rectilinear earthbag structures, alternate stacking the bags mason-style at the corners to interlock where the walls meet. Maintain a three-foot (.9 m) minimum distance from openings to corners (Fig. 5.10).

5.10: *Alternate overlapping bags at corners mason-style for extra strength. To increase the mass where these two Egyptian-style arches are located, we built the entire corner using 20-inch wide 100-lb bags.*

Tube Corners

Twist tight corners with tubes as prescribed in the section "Tips for Tubes" in Chapter 3. Tubes can also be interlocked mason-style, as you would with bags, or tubes can be extended to create exterior buttressed corners (Fig. 5.11).

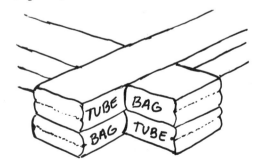

5.11: *Alternate tubes with bags to extend beyond the wall to create buttressed corners.*

Window and Door Openings

Provide ample solid wall in between openings. New Mexico's rammed earth code requires a minimum of three square feet (approx. 0.3 sq. m) of solid wall in between openings. Since rammed earth walls are typically 24 inches (60 cm) thick, there only needs to be 18 inches (45 cm) between the openings to fulfill the three square feet (0.3 sq. m) minimum requirement (1.5 ft. x 2 ft. = 3 sq. ft.). Because earthbags come in a variety of widths, either use wider bags or increase the distance between the openings to create three square feet (0.3 sq m.) of wall area. Another option is to incorporate buttressing between the openings. This allows for closer placement of windows, while still satisfying the three square foot (0.3 sq. m) minimum. It's probably a good idea, though, to make the space between openings at least 18 inches (45 cm) — the width of a typical working 100-lb. bag. Check out this illustration for ideas and a better understanding of what we are talking about (Fig. 5.12).

For curved walls, figure the minimum distance in between openings using the measurement on the inside surface of the wall that will still create three square feet (0.3 sq. m) of area in between openings. Keep in mind, however, this is for round, *vertical* walls, not corbelled dome walls (Fig. 5.13).

Generally, small openings like 2-feet by 2-feet (60 cm by 60 cm) square windows are compact enough to incorporate into a wall without jeopardizing the structure (see Chapter 8).

It is mostly large openings, four feet (1.2 m) and wider, that require ample mass between them. Windows with ample wall space between them do not allow for much direct solar gain, however. (For alternative ways to incorporate solar gain into an earthbag structure refer to Chapter 17). If creating lots of windows is a priority, consider designing a hefty post and beam or stud frame integrated into the earthbag wall (refer to Post and Beam section of this chapter).

When in doubt, we choose to overbuild rather than risk compromising the structure due to unforeseen circumstances.

Locking Row

After the fan bags have been installed, to complete an arch we like to lay two rows of either bags or tubes as a *locking row* over the top of the openings as a means of unifying the whole structure, as you would a conventional bond beam or top plate. The two locking rows also help distribute the weight of a

5.13: *For a curved wall, take measurement from interior of wall surface to calculate three-square-feet of wall in between openings.*

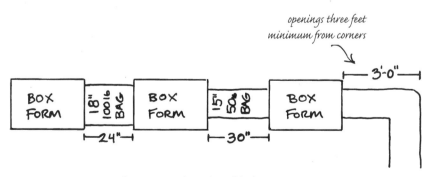

5.12: *Two ways to create three-square-feet of wall in between openings using different size bags.*

roof evenly over windows and doorways. This is particularly relevant when we are intending to build a structure without a conventional, poured concrete bond beam (Fig. 5.14). (For more information concerning bond beams, please refer to Chapter 9. For freestanding walls, one locking row has proven to be adequate).

5.14: *Two continuous locking rows of bags or tubes above the finished window or door openings.*

5.16: *The second row of bags lock down the Velcro plates and wrap around the posts.*

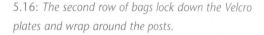

Designing Post and Beam

One of the most common strategies for getting a building permit in areas where earthen architecture is unfamiliar, or officials are heavily biased against it (usually from ignorance of the medium), is to incorporate a post and beam framework as the load-bearing structure, delegating the bag work as infill. This is contrary, of course, to all one's efforts directed at minimizing the use of lumber and energy intensive materials like cement and steel. At least it may help you get a house built, while introducing an alternative building method like earthbags.

One way to limit the use of lumber is by using small dimensional posts, like four-inch by four-inch (10 cm by 10 cm), set at the furthest distance allow-

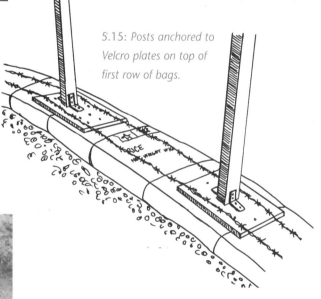

5.15: *Posts anchored to Velcro plates on top of first row of bags.*

able; about eight feet (2.4 m) apart. Here is an example of one system for installing a post directly on top of an earthbag stem wall for such reasons as building in a glass wall or infilling with strawbales (Fig. 5.15 & 5.16). Most post and beam structures require an engineer's stamp of approval and naturally they will want to beef up the top plate (bond beam) to cover their own butt. The bag work easily swallows up the posts by wrapping around them. The posts, set eight feet (2.4 m) apart, allow plenty of space to build window and door openings around forms, or add more posts to use as built-in window and door forms.

One advantage of using posts as the load bearing structure is that supporting the posts on concrete piers (instead of a continuous concrete foundation) permits you to do the bag work on top of a rubble-trench-and-stabilized or stone, stem wall foundation. If your project is not subject to the scrutiny of building code regulations, consider yourself luckier than if you had won the lottery. All those damn posts can sorely interrupt the flow of the wall system by breaking up corners with posts, etc. They also increase the cost per square foot and slow down the building process.

Perhaps one of the simplest and strongest post and beam configurations is a circle with 4-inch by 4-inch (10 cm by 10 cm) posts set at eight-foot (2.4 m) intervals on piers. Infill between the posts with earthbags set on a rubble trench foundation. Use forms to build arch openings. When the bag work is level with the height of the posts, build a faceted wood beam or top plate, or pour a concrete bond beam. Roof the structure according to taste. This recipe will at least help to lessen the use of wood, cement, and steel, while still meeting most building code requirements. It all depends on how much you are willing to compromise, depending on your personal point of view. Here are examples of what we are talking about when we speak of compromise.

Alison Kennedy wanted to build her house using earthbags, here in Moab, Utah. Although the building codes in Utah at that time allowed for load bearing adobe walls, the adobes had to be stabilized. Rather than adding the extra work involved in mixing cement into all the dirt, as well as killing the natural character of a living soil, she opted to install posts and beams in the form of precast concrete blocks (Fig. 5.17).

5.17: Alison Kennedy's code-approved earthbag home with concrete-block posts and beams.

Sarah Martin's addition onto the back of the Comb Ridge Trading Post in Bluff, Utah, made use of an existing foundation, set with posts, that had been installed by the previous owner. She and a crew of friends infilled all the bag work in two weekend parties that included food, libations, and copious amounts of laughter and merriment. The barbed wire was installed in a figure-eight pattern, weaving its way inside and outside each pole throughout the entire wall system (Fig. 5.18).

5.18: *Because the posts were fairly narrow, the bags swallowed them up as they conformed around them.*

Keep in mind that many of these design considerations are in direct response to the limitations imposed by existing building codes. Don't think of these restrictions as impedance to your desires to build with more resource-friendly materials. Rather, think of them as opportunities for creative problem solving. If you choose to work within the system, compromises will likely have to be made, but your building department will be making concessions, too. And you are providing them with the opportunity to learn new techniques and, just possibly, open their minds to something that works outside the status quo. Creative potential is contagious.

CHAPTER 6

Step-By-Step Flexible-Form Rammed Earth Technique, or *How to Turn a Bag of Dirt into a Precision Wall Building System*

This chapter is designed in an earthbag workshop format that demonstrates the Flexible-Form Rammed Earth technique and employs the FQSS stamp of approval. The Flexible-Form Rammed Earth technique is easy to learn and simple to perform. It is a pocketbook and resource-friendly building method that is enduring, beautiful, and widely accessible.

What we will present here is a step-by-step guide that will walk you through all the phases of building an earthbag wall. This same process can be applied to a garden, privacy, or retaining wall, as well as walls of a structure that is either rectilinear or curved. This, however, does not apply to dome walls. (A double curvature structure (dome) is discussed at length in Chapters 11 and 12). Learning the techniques in this chapter will aid you in better understanding how to build a dome when that time comes. This chapter is the logical place to begin that journey.

Foundations

A simple foundation can be a trench at least six inches (15.2 cm) wider than a *working bag's* width. This allows at least three inches (7.5 cm) on either side of the bag wall for plaster application and to discourage plant growth. The depth of the trench is dependent on frost level and soil conditions in your particular locale. Cut the trench and then fill with crushed rock. Place the coarsest material at the bottom. Moisten and tamp as you fill. Level the material within the trench (Fig. 6.1).

6.1: *Tamped gravel foundation.*

Foundations built on slopes can also be leveled with some of the bag work being buried as it rises from the gravel trench. A straight run of freestanding wall that is three feet (0.9 m) or more in height requires buttressing at every 12 feet (3.6 m) on center for lateral stability. Curved walls do not require buttressing, since the curve of the wall provides ample lateral support. Make sure to include any buttressing designs in the foundation plans (Fig. 6.2).

Depending on the type of foundation system you have chosen, begin the bag work either directly on the gravel trench or on top of the stem wall.

6.2: *Cut bank level gravel foundation.*

ROW 1

The Bag Stand

Start at one end of the foundation/stem wall. Place a bag on the stand so that the bottom of the bag rests slightly on the foundation. Toss in two cans of pre-moistened fill (Fig. 6.3).

Diddling

Reach down and push the corner in from the outside of the bag while packing the dirt up against it from the inside to secure it (Fig 6.4 & 6.5).

The results of diddling create a crisp solid edge that eliminates soft spots keeping the wall surface firm and smooth for plastering (Fig 6.6).

Hard-Assing

The first bag in any exposed end wall, corners, or buttress gets *hard-assed*. This procedure of pre-tamping the bottom of this bag results in hard blocky end walls that resist slumping when later rammed from above (Fig. 6.7).

6.7: *Hard-assing with a quarter pounder.*

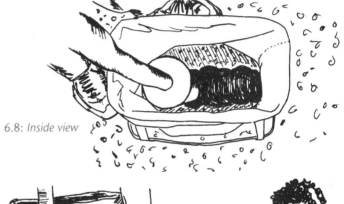

6.8: *Inside view*

Using a quarter pounder, tamp the soil straight down the middle of the bag. There's no need to tamp the full width of the bag. Just down the center will do (Fig. 6.8).

Add two cans of soil at a time and alternate tamping and filling until you have tamped the soil to 10-12 inches (25-30 cm) high (Fig 6.9).

Ditch the tamper and continue filling the bag until the soil is six inches (15 cm) from the top of the bag.

Remove the bag stand (Fig 6.10).

6.9

6.10

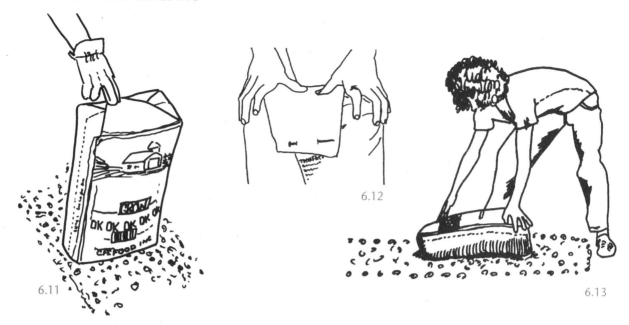

6.11

6.12

6.13

Pressing with your hands, firm the dirt in the top of the bag. Then fold the top of the bag snugly, like an envelope, (Fig. 6.11) and pin it shut with a nail (Fig. 6.12).

Lay the bag down onto the foundation two inches (5 cm) in from desired location of the finished wall, as the bag will expand when later tamped from above (Fig. 6.13).

The Second Bag

Start the second bag one bag's length from the first bag. Toss in two cans of dirt and diddle the corners. Using your fist, compact the dirt into the corners for the next several can loads. Continue filling until the dirt is six inches (15 cm) from the top seam of the bag (Fig. 6.14). Remove the bag stand and firmly press the soil. Fold and lower the folded end into the top end of the first bag. This will hold it shut without pinning. Place the bag *firmly* up against the bottom of the previous bag! (Fig. 6.15).

6.14

6.15

Now it's throw and go! Just fill, diddle, fold, and flop. Use a string line to guide straight runs, or mark lines on the foundation for curved walls (Fig 6.16).

Use *story poles* to create square corners and help maintain the level surface of the bag work as the wall height increases (Fig. 6.17).

For Moderate Curves

Use a wide brick or two-by-four to smack the end of the last bag into the desired contour (Fig. 6.18).

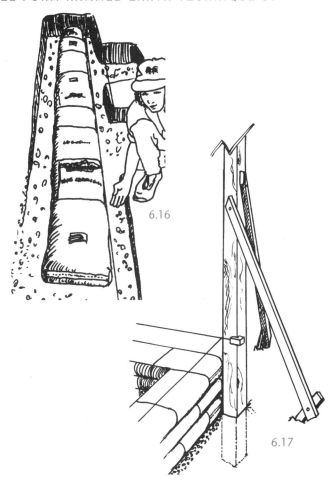

6.16

6.17

6.18

For Sharp Curves

To contour the bags for sharply curved walls, slope the soil in the top of the bag at an angle. Press firmly with your hands and then fold and lay the bag down. Sometimes pinning helps to create a little extra bag length (Fig. 6.19).

6.19:
Contoured bag.

Lock the Diddles

After an entire row is laid it is time to tamp. But wait! First *lock the diddles* on any hard-ass bags that may be exposed. Push a nail into the side of the bottom edge of the bag until it pokes out alongside the bottom seam of the bag (Fig. 6.20).

6.20

Catch a piece of the seam of the bag and gently wrench the nail around until the point is facing in the opposite direction. It will feel tight and may stress the fabric. Adjust tension to avoid tearing the bag (Fig. 6.21).

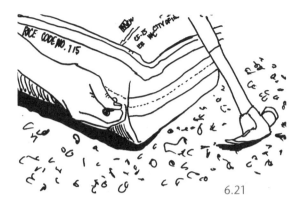

6.21

Hammer the nail in this opposite direction. It should feel tight enough that you will need a hammer to drive it in (Fig. 6.22). Do this in both bottom corners. That's it!

Locking the diddles on exposed hard-ass bags ensures block-like well-compacted ends that stay put (Fig. 6.23).

6.22

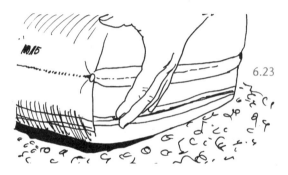

6.23

Tamping

OK, OK, OK; now we can tamp! Using a full pounder, begin by tamping down the center of the whole row. This forces the shoulders of the bags together to prevent shifting (Fig. 6.24).

After the whole row is first tamped down the center, then tamp from the center of the bags towards the outside. Continue to tamp the whole row until the compacted bags "ring." The sound will change from a thud to a smack. Average finished thickness for a tamped 50-lb. bag is five inches (12.5 cm) (Fig. 6.25).

Check the surface of the row for level. Tamp down the high spots. We like to stay within a half inch (1.25 cm) of level. This is pretty easy to do when starting with a level foundation.

6.24

6.25

Four-Point Barbed Wire

This is often referred to as Velcro mortar, hook and latch mortar, or "that #$@^&~%! pokey wire!" Have your barbed wire on a dispenser conveniently located at one end of the wall (Fig. 6.26). One at a time, lay out two parallel strands of barbed wire the entire length of the wall (Fig. 6.27).

6.26

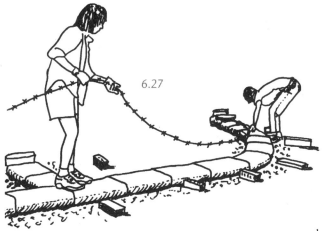

6.27

Have long bricks or rocks handy to hold the wire down as you lay it or tie on suspended brick weights (as described in Chapter 3). Incorporate any buttresses with the barbed wire (Fig. 6.28).

Cut wire long enough to over lap the ends (Fig. 6.29).

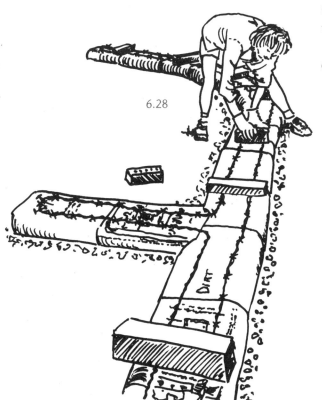

6.28

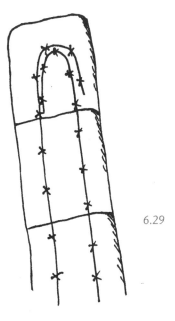

6.29

6.30

Tie Wires

If you intend to hang chicken wire or stucco mesh on the completed walls for stucco, add *tie wires* at this time. Loop a length of tie wire once around a strand of barbed wire to hold it in place. Make sure it is long enough to extend a few inches beyond the outside edges of the bags. Place one on each side about every 16-18 inches (40-45 cm). A good, clayey, natural earthen plaster will usually stick directly to the bags without the aid of chicken wire (Fig. 6.30).

ROW 2

Sliders

After the barbed wire is laid, a *slider* is placed under the bag stand (Fig. 6.31). Diddle and hard-ass the first bag in any exposed end wall or corner (Fig. 6.32). The slider aids in maneuvering the bag easily on top of the wire (Fig. 6.33).

6.31

6.32

6.33

Creating the Running Bond

Fill and fold the first bag shorter than the one underneath to create an overlapping or *running bond*. This will set the pattern for staggering the vertical seams for the whole row (Fig. 6.34).

6.34

Position the butt-end of the first bag about two inches (5 cm) in from the bag underneath. When tamped later from above, this bag will expand flush with the one below depending on the type of fill and how well it is tamped. Adjust according to your particular conditions (Fig. 6.35).

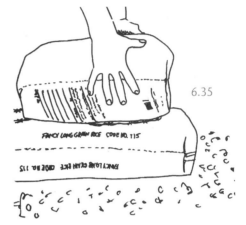

6.35

Remove the slider. Repeat the same procedure as in row 1, placing the slider under each bag being filled (Fig. 6.36).

Always install each bag firmly against the previous bag to create a tight, vertical seam (Fig. 6.37 & 6.38).

6.36

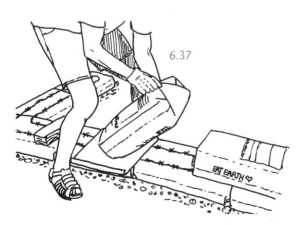

6.37

6.38

get a tight fit!

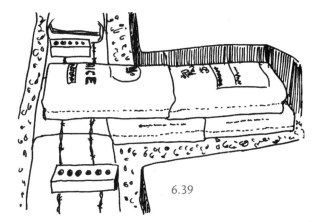

6.39

Integrate the Bags Where Walls or Buttresses Intersect (Fig. 6.39)

Note that both ends of these exposed end bags have been hard-assed, had their diddles locked, and have been laid two inches (5 cm) in from the exterior finished wall surface.

Their ends will extend out flush with the row below when tamped later from above (Fig. 6.40).

After the second row is tamped (and after each succeeding row), check again for level, noting any high and low spots. In a long run, it may be easier to check the level using a *water level*. (For directions on how to make and use a water level, refer to Appendix A) (Fig. 6.41).

Now is also the time to begin checking for *plumb* of the vertical surface. In order to maintain a wall that doesn't lean in or out (or both!), check the vertical surface with a level after each row is installed (Fig. 6.42).

6.40

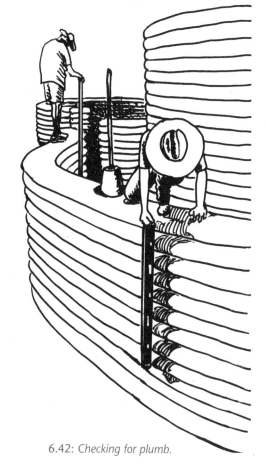

6.42: *Checking for plumb.*

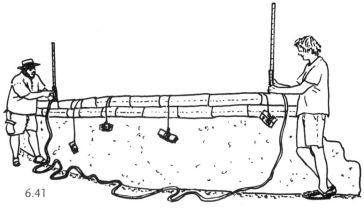

6.41

Door/Window Forms (Fig. 6.43)

To install a doorway, place a strong box or block form at the desired height and location. Level, square, and plumb the form. Construct door and window forms a couple of inches wider than the tamped width of the bag wall. This keeps the bags from wrapping around the edges of the forms, which would prevent the forms from being removed later. One way to create a form that is easy to work with (and remove later) is to use a *split box form* or *side wedge-box form* (Fig. 6.44). (See Chapter 2 for details on a variety of window and door forms).

Rather than build multiple box forms to accommodate the height of a door, consider using straw bales to elevate a single set of split box forms after the walls have been thoroughly compacted (Fig. 6.45).

Chicken Wire Cradles

It is helpful to wrap the bottom of the bags that are against the door and window forms with chicken wire. This provides a tight, grippy surface for the adhesion of stucco or earthen plaster. Extending the width of the wire beyond the width of the tamped bags provides a good anchor for additional sculpted adobe relief patterns and drip edges (Fig. 6.46).

6.43

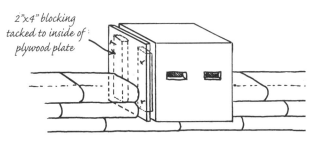

2"x4" blocking tacked to inside of plywood plate

6.44: *Doni suggests removing screws (or nails) from the plywood after a few rows of bags. Compression alone is enough to hold the side wedge board in place.*

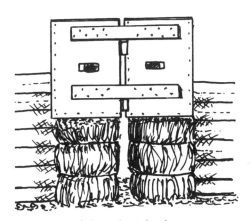

use straw bales to elevate box forms after walls have been compacted

6.45: *Note: This technique is used for plumb walls only - not for domes. Double curvature walls require constant compression until the dome is completed (See "Dynamics of a Dome" and "How We Built the Honey House.")*

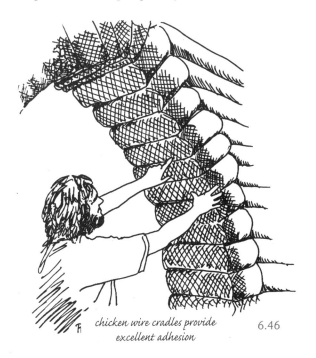

chicken wire cradles provide excellent adhesion
6.46

Cut the chicken wire about six inches (15 cm) wider than the tamped bag width and about 18 inches (45 cm) long. Bend one end of the wire about one-third of the length. Lay this shorter end on top of the barbed wire. Put your slider on top of this and then place the bag stand on top of the slider. Now you're ready for filling (Fig. 6.47).

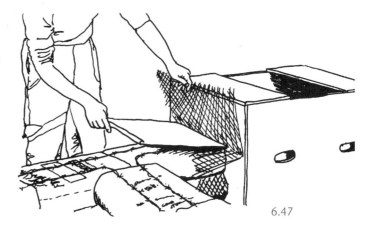

6.47

Hard-assing the bags alongside the door and window forms make them extra strong. Remember to diddle the corners, although it's usually not necessary to lock the diddles (Fig. 6.48).

Strip Anchors

Installing *strip anchors* during wall building will provide a solid wood attachment for bolting doorjambs, certain types of windows, cabinetry, and intersecting stud-frame walls (Fig. 6.49). Add strip anchors as often as every three to four rows or as needed. Four evenly spaced strip anchors on each side of a door opening is sufficient for bolting most doorjambs onto them.

6.48

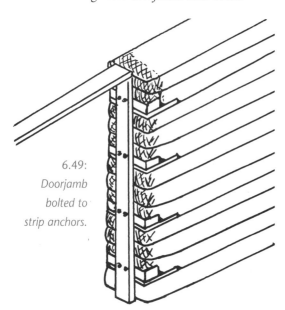

6.49:
*Doorjamb
bolted to
strip anchors.*

Push the two-by-four solid wood part of the anchor flush against the form. Velcro the plywood strip part of the anchor by hammering three-inch (7.5 cm) long nails through it into the tamped bag below. Make sure to pre-trim the strip anchors to conform to the shape of a curved wall (Fig. 6.50).

The barbed wire can now be laid on top of the strip anchor and the bag work can continue as usual. When installing a bag against forms directly on top of a strip anchor, it is usually unnecessary to hard-ass the bag as the strip anchor will take up too much space.

Teamwork

Follow all the steps previously outlined. As the walls get taller, the assistance of willing friends (or paid help) speeds the progress by sharing the work. For less down time, have several wheelbarrows loaded in a row where you are working, or have a third person recycling the wheelbarrows (Fig. 6.51).

Have enough people on hand when laying barbed wire high up. Someone guiding the reel of wire on the ground keeps it from tangling around the person up on the wall. Always use caution when working with barbed wire, doubly so when high up (Fig. 6.52).

6.50

6.51

6.52

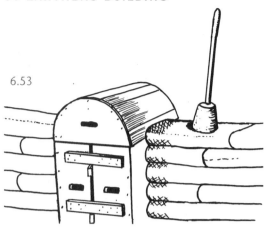

6.53

6.54

Arch Forms

We are now high enough to install our *arch forms!*
Place the arch form directly on top of the box form,
without the wedges. Continue the next row of bags as
an aid in locking the arch forms in position. Gently
half tamp the bags with equal force on both sides of
the arch form (Fig. 6.53).

6.55

When the arch form is secure, insert the wedges
on the front and back of the form and tap them in
until the forms are level and plumb. Tap wedges in
deep enough to create a good inch-wide (2.5 cm)
space between box and arch forms, to give ample
room for the arch form to be removed later.

When using just an arch without the box form,
set the arch form on a one-inch (2.5 cm) thick board
or piece of plywood the same width as the arch form
and the same depth as the bag wall. Drive some nails
into the board to Velcro it to the wall holding it in
place (Fig. 6.54).

The wedges will go in between the board and arch
form. The board will keep the wedges from digging into
the bags. You can now finish off by tamping the bags on
both sides of the form along with the rest of the wall.
This will secure the position of the arch (Fig. 6.55).

Fan Bags

These are the bags that begin the *springline* that creates
the arc of an arch. *Fan bags* are first tamped from the
inside, similar to a hard-assed bag. The main difference
is that a fan bag is tamped into a wedge shape and is
filled only 12 inches (30 cm) high. This height fan bag

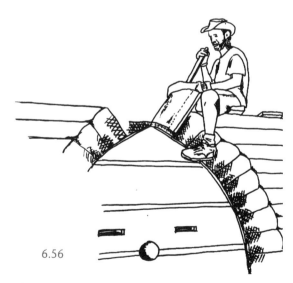

6.56

produces a strong, attractive framework around the
arch openings and ties in well with the barbed wire
mortar and surrounding bag work.

Hand Shaping a Fan Bag

Prepare a bag with diddled corners and chicken wire cradle. While building fan bags and hard-assed bags, place a slider on top of the chicken wire cradle to help keep the pokey stuff out of your way. Start tamping the dirt inside the bag (hard-assing). What we want is to firmly tamp the inside of this bag a little wider and wider towards the top to form the initial wedge shape (Fig. 6.56). When we reach 12 inches (30 cm) in height, snugly fold and pin the top closed.

Attach a string at the appropriate location along the base of your arch form (Fig. 6.57). The *springline* is where the curve of an arch begins sloping in. Use this line as a guide to align the angle of the fan bags surrounding the arch form. When a whole row of bags is ready to be tamped, tamp the fan bags to align with the angle of the stringline (Fig. 6.58).

Incorporate the barbed wire from the walls onto the surface of the fan bags to tie them into the wall system. The fan bags will continue to become more vertically oriented as the arch grows taller (Fig. 6.59). Keep an eye on maintaining the symmetry of the fan bags during construction, for balance, beauty, and structural integrity. (If using a *wedge box* to make fan bags, refer to Chapter 3 for simple directions on how to use this handy tool).

The Keystones for a Roman Arch

Continue constructing walls and fan bags together. When the space between the base of the fan bags narrows down to about eight inches (20 cm), it is time to add the *keystone bags* (Fig. 6.60). As a dirtbag rule of thumb, a narrower opening offers more structural resistance than a wider opening. When in doubt, add another row of bags, to reduce the space above the arch. The keystones provide a forceful downward and outward pressure that is met by the resistance of the walls on either side of the arch. This resistance is a form of buttressing (an arch is only as strong as its buttress). This resistance directs the compressive forces from above to the sides of the opening and down to the ground.

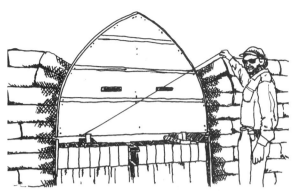

6.57: *Eight-point Egyptian arch springline.*

6.58: *Shaping the fan bags.*

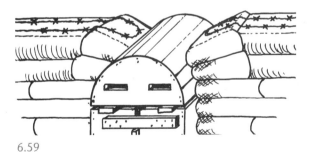

6.59

6.60

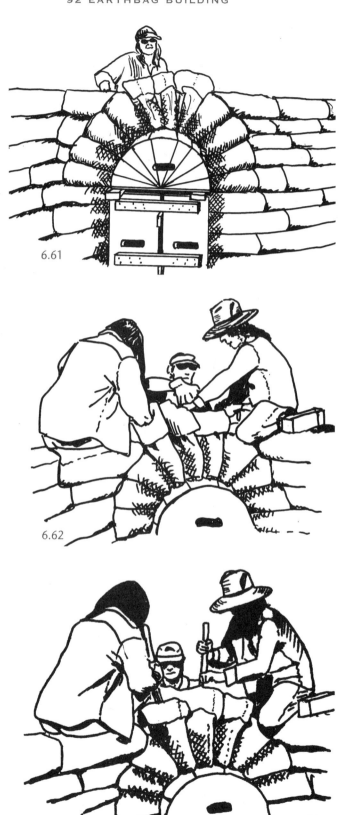

6.61

6.62

6.63

Installing the Keystones

Let's install our keystone bags. Hold off on laying the next row of barbed wire. While still on the ground, prepare three keystone bags; add two cans of dirt to each bag, diddle the corners, and tamp them on the inside with a quarter pounder. Cradle each one in chicken wire.

Neatly place all three bags side-by-side into the open wedge above the arch form (Fig. 6.61). After arranging the bags as symmetrically as possible, use a blunt stick or the handle of the quarter pounder to tamp the inside of each keystone bag.

Add two more cans of dirt and tamp the full interior width of each bag. Use firm, consistent pressure in all three bags. Treat them as one single unit. It helps to have two people tamping while a third provides the fill and quality control on how the bags are shaping from the best vantage point (Fig. 6.62).

Continue the process of tamping and filling, always adding the same amount of dirt to each bag. The bags will widen and widen until the open wedge space disappears (Fig. 6.63).

The keystone bags usually top off at the same 12-inch (30 cm) height as the other fan bags. That's it! Fold the tops of the bags over, cutting off any excess material and pin them closed tight with nails (Fig. 6.64).

6.64

Keystones for a Gothic Arch

Egyptian or Gothic style arches make a severe wedge-shaped space to fill. They can be very narrow at the bottom and easily require the full-tamping width of three keystone bags at the top. Continue bag work until a maximum of two to three inches (5-7.5 cm) of space remains before installing the keystone bags. Prepare and fill these keystone bags as described for a Roman arch keystone on pages 91 & 92 (Fig. 6.65).

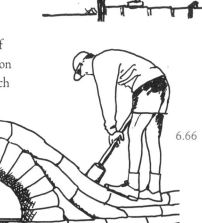

6.65

Locking Row

Before getting carried away with the excitement of removing the forms, we'd better add one more row of bags as a *locking row* to maintain downward pressure on the keystones. This will ensure the integrity of the arch after removal of the form (Fig. 6.66).

6.66

Lay the barbed wire over the top of the whole last row, including the keystones. Lay a final row of bags. Laying a *tube* (or *continuous bag*) is an excellent way to lock in the keystones, while integrating the arch with the rest of the wall. Bags or tubes will perform the same function. The wall will continue to become more stable as the fill material cures inside the bags (Fig. 6.67).

6.67

Removal of the Forms

Forms can be removed now or they can remain in the
wall as long as necessary. The longer they remain, the
more curing time to allow the wall to become stronger.
(Just something to think about, depending on design
and the quality of the earthen fill.) Tap out the wedges
with a hammer. The form is now free to drop down
and be pulled out (Fig. 6.68). To remove the wedges
from a *side wedge box form,* insert a pipe through the
space at either end of the blocking and knock out the
board at the opposite side (Fig. 6.69). The box form is
then free to be removed (Fig. 6.70).

Bask in your accomplishment.

6.68

6.69: To remove wedge blocks, insert a pipe through the
space at either end of blocking and knock out the block at
the opposite side.

6.70

SOME OF THE POSSIBLE CAUSES OF A FORM GETTING STUCK IN A WALL:

1. Screw or nail head on surface of arch sheathing caught on chicken wire.

2. Form was too short and bags wrapped around edge of the face of the form.

3. Form lacked sufficient internal bracing to prevent deflection during construction.

Fortunately for you, you followed directions and screwed the face of the form exposed to the outside so in case of a problem you could dismantle it, right?

6.71

CHAPTER 7

Electrical, Plumbing, Shelving, and Intersecting Walls: Making the Connection

Introduction

It is well worth reminding ourselves that we are building a whole house and not just the walls but everything that goes into the walls. The electrical sys-

tem, plumbing stub-ins, attachments for shelving, and intersecting walls are all installed during construction. The good news is, a lot of this work is done when the wall building is completed!

Electrical Installations

Make a drawing of an electrical plan locating where all outlets, switches, and runs will be placed. Pre-construct all the outlet and switch plates that will be needed (Fig. 7.2). Make a mark on the bag row below

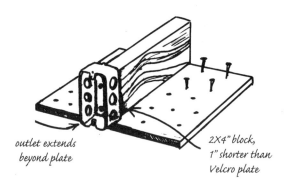

outlet extends
beyond plate

2X4" block,
1" shorter than
Velcro plate

7.2: *Screw elecrical box to 2"X4" block. Position plate on wall so that outlet will extend far enough to become flush with plaster later on.*

7.3: *Electrical outlets and conduit installed, barbed wire nailed to Velcro plates; ready to continue bag work.*

where the outlets and switches are to be installed. Position and Velcro the plates into place when you get to those points.

Electrical Conduit

Some electrical codes may require UL-approved rigid metal conduit or flexible, "armored" conduit to run the wire through. If you are using conduit, install it now. Lay out your barbed wire and continue wall building (Fig. 7.3). The disadvantage of rigid conduit is that it is hard to use it for vertical runs and for bending around windows and doors. Conduit can be run vertically on the face of the bag walls but will require extra plaster to conceal it later. Flexible conduits like *flex metal, water tight,* or plastic *Smerf* pipe is easier to bend and snake around bags. For horizontal runs, either pipe or conduit can be cinched tight in between two rows on the surface of the wall with tie wires or by

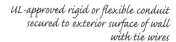

UL-approved rigid or flexible conduit
secured to exterior surface of wall
with tie wires

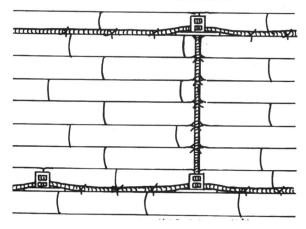

7.4: *While the earth is still green, you can dimple a channel with a hammer to sink the conduit flush with the wall.*

swinging the suspended brick weight lines around them. Flexible types of UL-approved electrical conduit are fairly pricey. They average about 50-60 percent more than rigid metal conduit (Fig. 7.4).

UF Cable

UF cable is designed to be buried in the ground for outdoor applications. It is both waterproof and crush resistant, unlike standard Romex cable, which is neither. New Mexico Adobe Code approves the use of UF cable buried in the center of the wall. UF cable is the least expensive method. The cable is connected from outlet to outlet. Any wire that is destined for vertical runs can be snaked up in between the bags.

Because we are also laying barbed wire in between our rows, it is advisable to provide some sort of protection for the cable, or carefully tack down both the cable and the barbed wire with nails driven at an angle to secure them until the bags are laid.

Power Entry

Remember to install Velcro plates (strip anchors) where the electrical panel will be located. Plan on installing a one-and-one-quarter- inch (3.125 cm) conduit or plumbing pipe through the foundation to stub-in the entry of electrical wire from a buried origin. If power is being supplied from overhead, place enough strip anchors facing the exterior surface of the wall to provide attachment for the electric entry "masthead."

Plumbing

As with the electrical system, have a plumbing design plan figured out before the walls are started. Incoming water and outgoing drains will likely come in under the foundation or through the stem wall, depending on frost level and the type of waste management system. The plumbing arrangements for an earthbag wall are pretty much the same as they would be for any other type of construction method. Horizontal pipes can be laid in between the surface of the rows of bags and cinched tight with the wires, using the same strategy discussed for electrical conduit. Horizontal plumbing pipe can also be buried

beneath an adobe floor. Likewise, the vertical runs can be channeled into the bags with a hammer, to recess the pipe as much as possible into the wall. Having access to pipes eventually pays off when there is a leak (Fig 7.6).

Most plumbing is hidden by counters or runs along the backside of built-in sculpted adobe benches. Cob and earth plaster can hide almost anything and, if a leak does occur, it will be noticeable and easy to access. Adobe and rammed earth builders sometimes run major plumbing systems, such as a shower stall system, through an attached or intersecting frame wall.

People are developing many effective, ecologically friendly, alternative wastewater systems. It is worth exploring other options to conventional sewer and septic tanks. Introducing them all is beyond the scope of this book. Search the Internet for permaculture sites and natural gray and black water management systems, as well as the Resource Guide in this book.

Intersecting Stud Frame Walls

Earthbags make wonderful, sound-dampening intersecting walls and room dividers. If, however, space is at a premium, small spaces like closets and bathrooms will take up less floor space if built out of thinner wall

install velcro plates to attach plumbing pipe

7.6: Install a larger diameter pipe underneath foundation as a sleeve for incoming and outgoing plumbing.

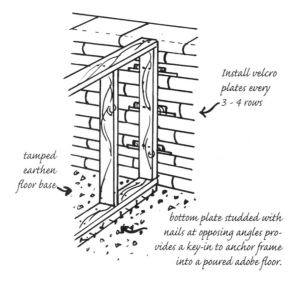

Install velcro
plates every
3 - 4 rows

tamped
earthen
floor base

bottom plate studded with
nails at opposing angles provides a key-in to anchor frame
into a poured adobe floor.

7.7: Intersecting stud frame wall lag-screwed
to strip anchors.

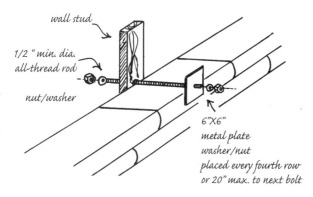

wall stud

1/2 " min. dia.
all-thread rod

nut/washer

6"X6"
metal plate
washer/nut
placed every fourth row
or 20" max. to next bolt

7.8: Stud frame wall attached with all-thread and bolted
through an earthbag wall.

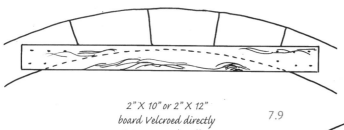

2" X 10" or 2" X 12"
board Velcroed directly
into a curved wall to
create a shelf.
Pay attention during
installation to assure
maintenance of level

7.9

materials like wood. The following are a couple of
techniques for attaching intersecting stud frame walls
to earthbag walls:

- Strip anchors set into the earthbag wall with a
  stud frame wall bolted to it with lag screws
  (Fig. 7.7).
- A square plate with a washer and bolt attached
  to it. Make sure the square plate is large
  enough to span at least half of the width of
  the bags below and above the bolt (Fig. 7.8).
- Our preference is to use narrower bags or
  tubes for interior walls, or hand sculpt walls
  out of cob or wattle and daub. The possibilities
  for alternatives to building with wood are lim-
  itless, while there *are* limits to the availability of
  wood. If an entire wall is to contain several
  large closely-spaced windows to take advantage of solar gain, it is better to build this wall
  of wood.

Shelving, Cabinets, and Stair Attachments

Plan where counters, cabinets, stairs, and shelving will
go. As the walls go up, so too will the Velcro plates and
strip anchors for anchoring this built-in furniture. An
assortment of Velcro plates with two-by-four blocking attached to them in various configurations are
needed for attaching cabinets flush against the wall.

Set the Velcro plates or strip anchors before
installing the next row of barbed wire. Nail the barbed
wire over the Velcro plate and continue the run along
the rest of the wall. "U-nails," used for attaching wire
to wooden fence posts, work great for attaching
barbed wire onto Velcro plates. To put a straight
shelf in a curved wall, Velcro large, 2- by-12-inch (5
cm by 30 cm) or wider, rough sawn dimensional lum-

7.10: Strip anchors follow the contour of a curved
wall for later attachment of cabinets.

ber, directly into the wall (Fig. 7.9). Another option is to run short lengths of strip anchors that follow the curve (Fig. 7.10).

Velcro Shelf Brackets

Installing Velcro shelf brackets on a curved wall is also possible. Instead of using shelf boards, try sheathing the brackets with willow saplings or green bamboo while they're still flexible. Leave them exposed, or pour a thick adobe veneer to make a smooth countertop (Fig. 7.11 & 7.12).

7.12: *Shelf brackets on a curved wall sheathed with bamboo and/or saplings.*

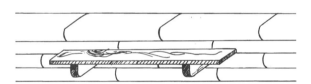

tamp wall as level as possible, position and Velcro shelf brackets. Lay barbed wire and next row of bags, tamp, check level again. shim if necessary

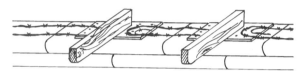

built-in shelves are great on exterior walls protected by porch roofs, too! and are an excellent source of scaffolding during construction

7.11: *Velcro shelf brackets: Before and after shelf installation*

Stairs

The same technique can be used for built-in stairs by staggering hefty timbers or large 4- by-10-inch (10 cm by 25 cm) planking. Attach to Velcro plates, spacing the height and length properly to create built-in steps (Fig. 7.13).

7.13: *Stagger plank steps during construction, seated on Velcro plates.*

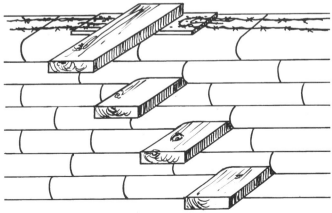

Nichos

Nichos are cavities recessed into a wall. Nichos are designed to provide shelving without protruding into the living space. We build deep nichos into an earthbag wall the same way we do arched windows. Shelving is installed, supported by a thick layer of plaster after the earthbag work is finished (Fig. 7.14). Shallow nichos can be carved into the wall with a hatchet.

Non-wood shelving can be sculpted from adobe, rich with straw, over temporary arch forms, or bamboo, carrizo, long bones, long stone, old pipe, pieces of rebar, or whatever can serve as an extension from the wall to sculpt mud around. Use nails, sticks, or bones as a key-in for cob. Begin sculpting around the forms at floor level, or lay up a few rows of earthbags. Set forms on top of these arch openings. Sculpt with an adobe-cob mix over the forms. Fill in the gaps between the forms to make a level surface (shelf). When the mud is set up, remove forms and place on top to sculpt the next set of shelves (Fig. 7.15).

secure shelves with plaster
construct deep set shelving using forms — just like windows
— cover exterior of back wall with heavy gauge chicken wire
— reinforce with cob or straw/clay, etc.
seal with plaster

7.14: *Nichos with plastered-in shelving for either interior or exterior walls.*

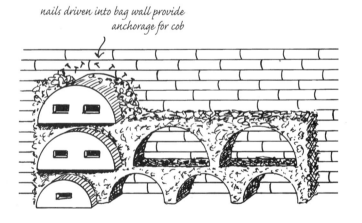

nails driven into bag wall provide anchorage for cob

7.15: *While we're at it, we might as well use our arch forms to sculpt furniture out of cob.*

Attaching into an Earthbag Wall After Building is Completed

After a few weeks to months of curing, a good quality rammed earth soil will hold nails pretty well. For more serious anchorage, drill holes with a masonry bit and tap in a plastic or metal sleeve (bolt anchor) designed for concrete and masonry walls. If the soil is poor quality, squeeze some concrete glue into the hole prior to tapping in the sleeve, let the glue cure, then screw or bolt into it. For anchoring heavy stuff, stick with the Velcro plate strategy.

CHAPTER 8

Lintel, Window, and Door Installation

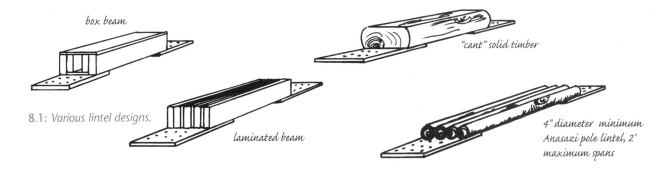

box beam

"cant" solid timber

laminated beam

4" diameter minimum
Anasazi pole lintel, 2'
maximum spans

8.1: *Various lintel designs.*

Lintels

Lintels are to earthen architecture what headers are to stud frame walls. A lintel is a sturdy beam that spans the space above a door or window opening that bears the weight of a roof or second story. Traditionally, they were made from large dimension lumber. These days, lintels are often built up from laminated small-dimensional lumber or constructed into a box beam. Pallets are an excellent resource for Velcro plates and for making laminated lintels. Whatever the design, our focus is on how to anchor a lintel to an earthbag wall (Fig. 8.1).

On average, lintels need to be at least three-quarters the width of the wall, and extend past the opening to rest on the wall a minimum of 12 inches (30 cm) on either side. Our approach is to attach Velcro plates to the underside of each end of the lintel to extend another eight inches (20 cm) beyond the lintel (Fig. 8.2). The Velcro plate provides a pad that protects the wall from the point of contact from the lintel, while distributing the weight over a

8.2: *An example of a lintel pre-attached to Velcro plates that extend another eight inches beyond the ends of the lintel.*

103

broader area. The Velcro plate also anchors the lintel during construction.

Structural dimensions for load bearing and shear-strength change with the length of the opening being spanned. The bigger the opening, the beefier the lintel must be. Check on structural requirements appro-

priate for your design. When designing the dimensions of a lintel, consider rounding off the thickness (or height) so that, including the thickness of the Velcro plate, it is equal to the thickness of the bags being used. This will make it easier to maintain the level of the bag wall (Fig. 8.3).

For a narrow opening, of two to three feet (60-90 cm) maximum, a minimum five-inch (12.5 cm) thick lintel is needed. For wide spans, of three to four feet (90-120 cm), a ten-inch (25 cm) thick lintel is called for. According to New Mexico Adobe Codes, 12-inch (30 cm) tall lintels are advised for spans over five feet (1.5 m). Occasionally the lintel and the bag wall may end up at different levels. You can either over- or under-fill the bags, or, if the lintel is lower than the bags, shim it with wood or throw a layer of cob on top to bring it up to level. Wait until the cob sets up some before continuing the bag work. To further secure a lintel, we like to lay a minimum of two rows of bags over them. The extra anchorage is particularly advisable when preparing the walls for a conventional roof system without a conventional bond beam (see Chapter 9).

8.3 (above): *In designs with multiple windows, they can share Velcro plates or the lintel can span across the top of all the windows.*

Window Installation

Windows can be installed onto a wood frame that is attached to strip anchors, directly onto the strip anchors, or shimmed and set with plaster alone. It is a matter of personal preference and how accurate the rough openings turn out.

Earthbag walls are an ideal medium for sculpted windowsills. We like to seat the window up tight against the underside of the opening, leaving us more space below to slope a stabilized earthen, mortared stone, wood, or poured concrete windowsill. Be sure to consider what design features are planned for the windows so that the box forms can be built to accommodate a thick, sloping sill (Fig. 8.4).

8.4: *The beauty of an arched window is that the shape is both attractive and structural. We choose the Roman arch (semi-circle) for most of our habitats because it is the easiest shape to fit operating windows into.*

We are aware of the environmental impacts of using both wood and vinyl windows. More and more factory wood windows are using composite wood products processed with a multitude of synthetic chemicals. The production of vinyl poses numerous environmental health hazards. However, as wood becomes more scarce and more expensive, even manufacturers of wood windows are including vinyl components. The purist may have to build his or her own or rebuild salvaged wood windows to insure a natural product.

Many bizarre yet practical solutions are being created and implemented to counter our toxic habits. Teruo Higa's book, *An Earth Saving Revolution II*, describes the benefits of using "effective micro-organisms" (EMs) to create a new breed of safe biodegradable vinyl that will decompose readily when buried. We are in an accelerated state of transition that we find both scary and exciting. Magic is afoot! Doing what we enjoy makes the world a better place, rather than having to make the world a better place in order to enjoy. Conclusion: use what is available. Ask the universe for solutions. Follow the path with heart. Breathe.

8.5: *Manufactured vinyl window with homemade wood sill and a half-round glass protected with a hose gasket. Both windows are secured with shims until sealed with plaster.*

Installing a Vinyl Window into an Earthbag Arch Opening (Fig. 8.5)

The appeal of vinyl is its low cost, compatibility with mud, efficient seal, thermopane glass, and fitted screen! Secure the window in the rough opening with shims, making sure to check for level and plumb. Trim the exterior flange on the pre-manufactured window if it is too wide (hand pruning shears work great for this). Install a sloped two-by-four sill on top of the window. Add a stop along the top of this sill to rest the half-round glass flush against (Fig. 8.6 & 8.7).

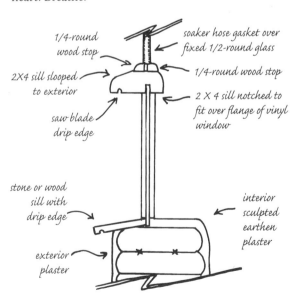

1/4-round wood stop

soaker hose gasket over fixed 1/2-round glass

2X4 sill slooped to exterior

1/4-round wood stop

saw blade drip edge

2 X 4 sill notched to fit over flange of vinyl window

stone or wood sill with drip edge

interior sculpted earthen plaster

exterior plaster

8.6: (Above) *Side view detail of vinyl window installation.*
8.7: (Right) *Split a length of five-eights-inch (1.56 cm) soaker hose as a gasket over a one-quarter-inch thick plate or thermoplane glass that has been pre-cut in a half-round shape.*

8.8 (left): *Install the half-round pane of glass up against the stop that was installed previously on top of the vinyl window. Plumb with a level and secure with shims.*

Mud it into place, leaving the shims imbedded in the mud until cured. Once it is cured, remove the shims and fill the gaps with more mud. Congratulations! You now have a beautifully installed, operable, finished window (Fig. 8.8).

Simple, homemade, operable windows can be made with a minimum of materials. Consider plastering in a piece of fixed glass (with a hose gasket seal) on top of a small, operable wooden door or awning for ventilation (Fig. 8.10a & b). Glass bottles mortared with mud into an opening are another way to let light into a building (Fig. 8.11). Another innovative

8.10a (above): *Cabinet style wood doors*

8.10b (below): *Fixed glass with awning-style vented wood opening*

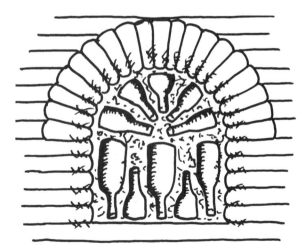

8.11: *Glass bottles, mortared in with adobe, provide insulation while giving ambient light.*

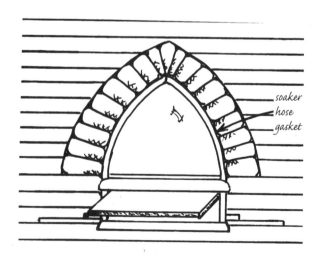

soaker
hose
gasket

8.10a and b: *Two examples of arched, fixed glass with wooden doors or awning set below for ventilation.*

window idea is to install a car windshield (under a lintel) into the wall. Windshields are strongly built and many have the added advantage of being pre-curved for custom fitting into a round wall. As in the other examples, seal it with a hose gasket and mud it into place with cob (Fig. 8.12).

Door Installation

Doors are easy! Shim, plumb, and level the doorjamb. Screw the jamb into the pre-placed strip anchors, just as you would for a stud frame wall, and plaster up to the jamb. Or build a framed rough opening and attach the jamb to that (Fig. 8.13a & b).

Note in Figure 8.14 (on the next page) the sculpted jamb detail above the door. Instead of using wood to create a curved jamb, we cinched saplings up tight in the curved wall with tie wires through the chicken wire cradles, and sculpted over them with mud plaster (Fig. 8.14).

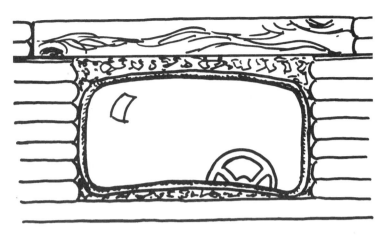

8.12: Car windshield sealed with a hose gasket, secured with adobe/cob under a wood lintel. What could be better than to view life through a Chevy?

8.13a

fixed glass arch over a wood sill plate

8.13b

8.13a & b: Two types of doors: One with lintel, one with fixed glass arch above.

Designing for Future Additions

It is possible to saw through an earthbag wall using masonry bits and to knock out chunks of the wall with a sledgehammer. The barbed wire will, of course, need to be cut. The opening, however, will be rough and there won't be any strip anchors in place in which to bolt a doorjamb (Fig. 8.15).

It's easiest to install a doorway during the initial construction and fill it temporarily with stacked straw bales and light straw/clay, or cob, or earthbags filled with a loose material or dry sand. Protected by an earthen plaster, you'll have your hole for a future addition, and a wall for current living. Stay in the moment, but think ahead! (Fig. 8.16).

8.14: *Arched doors need to be hinged near the edge of the opening so that they can open without banging into the arch above them.*

8.16: *Straw bales stacked in an arched doorway.*

MARA CRANIC

8.15

CHAPTER 9

Roof Systems

There are a zillion styles and methods of building a roof, many of which can be adapted to sit on earthbag walls. Our job is to show some techniques with which to anchor the roof to the walls and share roof styles that we feel complement the earthbag system. We are big fans of Native American architecture as well as vernacular architecture worldwide. All we need to do is look at how indigenous peoples built their homes to suit their environments to see what design features we, too, would find practical. Native American styles vary dramatically from earth lodges and tipis of the Great Plains to the majestic timber and plank houses of the Pacific Northwest.

9.1: *A colorful variety of asphalt shingles turns this roof into a work of art.*

The most obvious consideration is designing a roof that protects the walls appropriately for the climate. Longer eaves are called for in a wetter climate. Dry climates can take advantage of the use of *parapets* and *vigas* (log beams) commonly seen in the Southwest. Moist climates are natural watering systems for a *living roof*, while in a dry climate the roof can be used to harvest precious rainwater.

Bond Beams (Fig. 9.2)

Due to historic use, building codes in the Southwestern United States include structural design standards that pertain specifically to earthen architecture, many of which we have adopted to earthbag construction. Modern adobe and rammed earth buildings require a *continuous bond beam* built of either wood or concrete installed on the top of the finished earthen wall. The bond beam acts as a *tension ring* that ties all the walls together into one monolithic frame.

9.2: The bond beam must be continuous, covering the full perimeter of the wall.

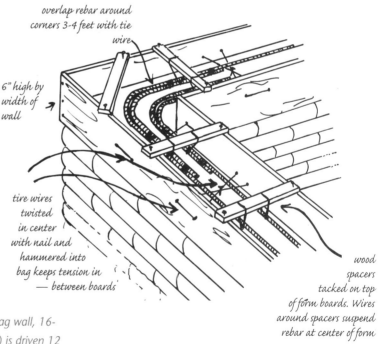

overlap rebar around corners 3-4 feet with tie wire

6" high by width of wall

tire wires twisted in center with nail and hammered into bag keeps tension in — between boards

wood spacers tacked on top of form boards. Wires around spacers suspend rebar at center of form

9.3: To anchor the bond beam to an earthbag wall, 16-inch (40 cm) long, #4 reinforcing bar (rebar) is driven 12 inches (30 cm) into the green (uncured), tamped earthbags at a maximum 20 degree angle, at least 4 inches (10 cm) in from the outer edge of the bag, and staggered at 24-inch (60 cm) intervals.

9.4: A minimum of two continuous #4 steel reinforcing bars are suspended in the form to provide tensile strength for the concrete.

A concrete bond beam is like a foundation on top of your walls. Codes vary from state to state, but typical dimensions are six inches (15 cm) high by the width of the wall. Most concrete bond beams are poured into wood forms that have been built on top of the walls for this purpose and then removed after the concrete has cured. The bond beam is secured to the wall by the opposing angles of rebar, thereby preventing uplift of the roof caused by high winds (Fig. 9.3 & 9.4).

Bond beams can also be built of wood in the same dimensions as a concrete bond, using either massive solid timbers or laminated lumber. A version of the adobe and rammed earth building codes can be obtained from the *Adobe Builder*, an architectural trade journal that publishes a book for adobe codes and one for rammed earth codes (Fig. 9.5, 9.6 & 9.7). (Check the Resource Guide in the back of this book).

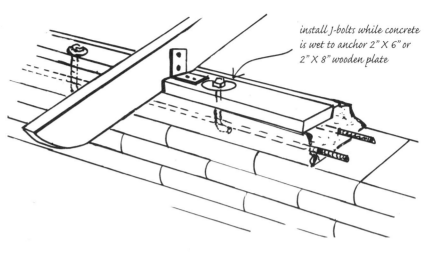

install J-bolts while concrete is wet to anchor 2" X 6" or 2" X 8" wooden plate

9.5: The bond beam also provides an anchor for attaching and distributing the individual weight of the roof members, whether they are rafters, trusses, or logs. In some cases it may double as lintels for the window and door openings.

9.6: A 2 x 6-inch (5 x 15 cm) woooden frame used as the concrete bond beam formwork can be left in place, and does double-duty as an attachment for the roof rafters.

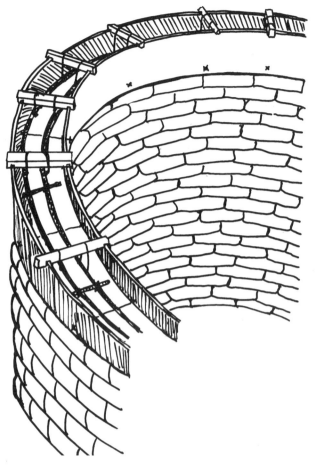

9.7: *For curved walls, thin, flexible Masonite can be used as a form.*

An alternative to the heavy wood bond beam prescribed by code is a light wood *ladder roof plate* anchored to the top of the wall with *poly strapping* cinched tight with a *tensioner device* (Fig. 9.8). (Refer to "Velcro Plates" in this chapter for more information on poly strapping and tensioners).

Cost Considerations

The cost of a roof can equal or exceed the cost per square foot of all of the exterior walls combined. This is particularly the case when building the roof with conventional building products. The easiest way to reduce roofing costs is to build modestly sized structures with relatively short spans, using as much minimally processed materials as possible. Since tensile strength is built into every row of an earthbag structure with 4-point barbed wire, the integrity of the entire structure manages stress with less dependence on a bond beam. When building with tubes, the tensile strength is further increased.

Bond beams have their place in high earthquake areas. Bond beams should be seriously considered for large structures and heavy compression style roofs. Consider also that many building techniques throughout the world have successfully survived the harshest environmental impacts for centuries before the introduction of the concrete bond beam. In earthbag construction, we rely on careful design, precision, and mass to hold everything together. So let's explore our other fun, quick, simple solutions.

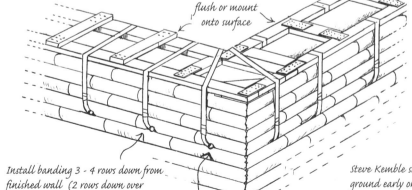

set 2" x 4"s flush or mount onto surface

Install banding 3 - 4 rows down from finished wall (2 rows down over window and door openings). If soil quality is poor, run banding through sections of 3/4 - 1" irrigation tubing

9.8: *Light wood ladder roof plate cinched to wall with woven poly banding (strapping).*

Steve Kemble says, "build ladder in sections on ground early on using the foundation as a template". Connect sections on top of wall with metal truss plates, or screw diagnal plywood plates onto corners.

Introduction to Alternative Roof Systems Without Bond Beams (Fig. 9.9)

Traditional earthen architecture was built without concrete, steel, or fossil fuel products. We feel that, in most cases, concrete and heavy wood bond beams are an unnecessary use of money and resources. (If you are considering building a roof system onto an earthbag structure without a continuous bond beam, please review Chapter 5). As a review, and to prepare for building a roof without a bond beam, these structural features should be taken into account.

We would not advise putting a heavy compression style roof on earthbag walls without a concrete or wood bond beam. (Domes are a separate category, presented in Chapter 11).

All of these roof systems direct the weight of the roof straight down onto the walls. When we've met all the structural safety features we're ready to put up our roof (see Fig. 9.10 on following page).

Velcro Plates

We use Velcro plates for attaching just about anything. For roofing, they work great as a platform on which to secure rafters, vigas, or trusses. Their main function is to distribute the weight of the individual roof members to keep it from digging into the earthbag wall. The plate is attached to the roof member prior to being lifted onto the wall. When the roof member is correctly positioned, the plate is Velcroed into place with three-inch (7.5 cm) long galvanized nails.

The Velcro plate alone is not enough to secure the rafter. The rafter needs to be anchored firmly to the wall to prevent uplift from wind blowing up underneath the eaves. Rafters can be secured by banding or strapping that has been installed three to four rows below the top row of bags. The strapping itself is installed during construction (Fig. 9.11a & b on page 115).

In addition to the tie-down method, another row of bags can be placed atop the Velcro plates in between the rafters. Besides adding additional anchoring to the Velcro plate system, the bags fill the spaces between the rafters up to the level of the roof itself. The bags

9.9: *This Anasazi structure in Chaco Canyon, New Mexico, shows the original vigas and latillas sitting on stone and mud walls, still intact after 1,000 years. Will a modern tract house ever get to make this claim?*

CHECKLIST FOR SUCCESS

Combine all of these features:

• Provide adequate solid wall in between openings.

• Keep the openings relatively small.

• Integrate interior walls and/or buttressing.

• Add two locking rows.

• Keep roof spans short or build internal supporting walls or post and beam structure.

• Choose a roof design that exerts pressure downward instead of outward.

also provide each rafter with lateral support. Any of these tie-down systems can be used for trusses and vigas as well (Fig. 9.12).

9.10: *Examples of roof styles suitable for earthbag buildings.*

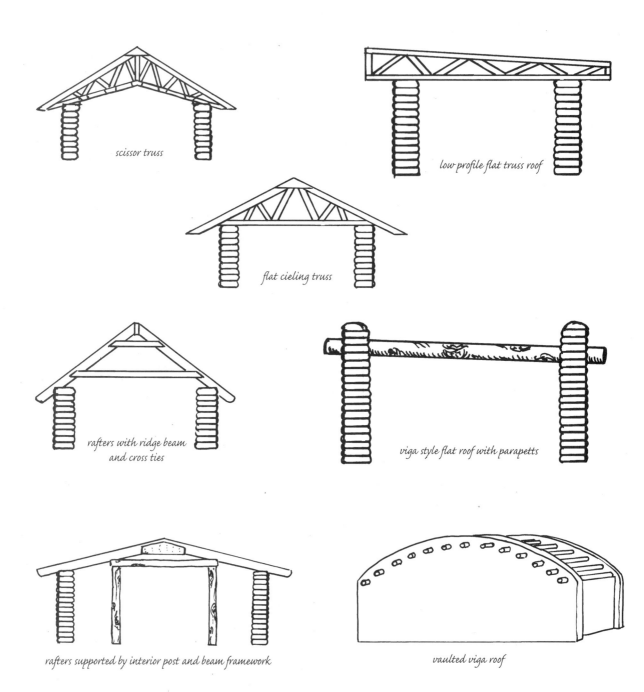

scissor truss

low profile flat truss roof

flat cieling truss

rafters with ridge beam and cross ties

viga style flat roof with parapetts

rafters supported by interior post and beam framework

vaulted viga roof

9.11a & b:
*Attaching rafters
or trusses to
earthbag walls.*

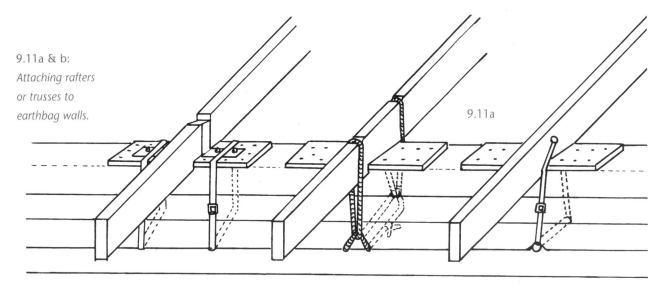

9.11a

rafter bolted to Velcro plate with steel
bracket, Velcro plate cinched tight with
banding

rope over notched rafter cinched
through exposed ends of a loop of
knotted rope

banding cinched through hole in rafter
and 3/4" poly pipe set into wall

9.11b

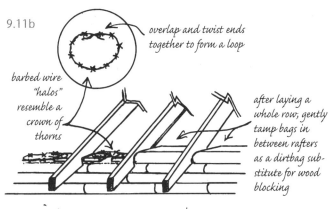

overlap and twist ends
together to form a loop

barbed wire
"halos"
resemble a
crown of
thorns

after laying a
whole row, gently
tamp bags in
between rafters
as a dirtbag sub-
stitute for wood
blocking

*nailing halos of barbed wire in between rafters
creates a mini tension ring*

Vigas

An alternative to strapping vigas is to anchor them
with rebar through pre-drilled holes at opposing
angles (Fig. 9.13a & b).

Rafters and vigas designed for low-pitched, flat
roofs can be anchored from above by building a *parapet.* Old adobe buildings used this strategy as a means
to anchor vigas and prevent uplift by placing weight
on top of the logs. This technique is still used in
many countries without the addition of concrete bond
beams (Fig. 9.14).

9.12: *Assembly of banding (strapping) tools.
Tensioner, 1,300 feet (390 m) of strapping (600 lb
capacity), crimps, and crimping tool cost about $130.
Compare this to the cost of a concrete beam.*

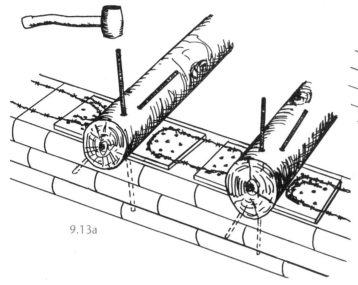

9.13a

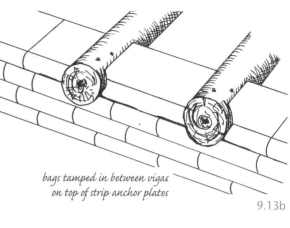

*bags tamped in between vigas
on top of strip anchor plates*

9.13b

9.13a: *Attach viga to velcro plate. Anchor plate to wall with 3-inch long galvanized nails. Pre-drill holes at a 20-degree angle through viga and velcro plate. Drive rebar 12 inches (30 cm) deep into bag wall. If the earth has cured, pre-drill into the earth using a bit one size smaller than the* size of the rebar. Pour concrete glue into the hole and tap in rebar (this will help keep the earth from fracturing). Nail halos of barbed wire onto velcro plates in between vigas.

9.13b: *Lay in the next row of earthbags.*

9.14:
Flat roof with parapet walls and canales.

*wood box canale lined
with galvanized sheet
metal*

9.15:
Southwestern style canales.

Roofs completely enclosed with parapets need to have *canales* built into them. Canales are short gutter spouts located at the low end of a shallow-pitched roof that directs water away from the walls of the building and through the parapet (Fig. 9.14 & 9.15).

Vaulted Viga Roof

One design we have been playing with is a *vaulted viga* roof that uses parapets on the two sides and eaves on the ends. It is our version of a low-tech organic substitute for a singlewide mobile home. Its long, narrow shape provides short spans, while interior intersecting walls and external buttressing add stability and charm (Fig. 9.16 & 9.29 on page 121).

9.16: *Stylized vaulted viga with living roof.*

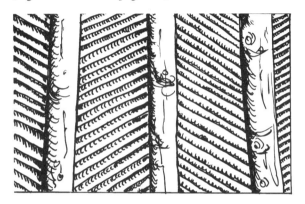

In a dry, moderate climate, the roof can be insulated from the outside with a topcoat of straw bales, or seeded into a living roof in a wet climate (see Figure 9.30 for a detail of a straw-bale-insulated vaulted viga roof). The vaulted viga roof is left exposed on the interior. Sheathe the top of the logs with long flexible boards, latillas (short, sturdy poles), or dismantled pallets (Fig. 9.17)

9.17 (above): *Vigas set parallel to each other with latillas set in an alternating diagonal pattern.*

9.18 (below): *A great example of the earthbag kiva is Penny Pennell's 36-foot (10.8 m) diameter home that uses a 6-sided post and beam interior support structure for the 20-foot (6 m) long vigas to rest on. (Credit: Penny Pennell)*

Special Considerations for Round Houses

Roofs for round buildings tend to be a little trickier because of all the angles, but they make up for it in beauty and aerodynamics. If you want to build a

9.19 (left): *Penny had a framework of dimensional lumber installed on top of the vigas, sheathed in plywood and insulated underneath.*

9.20 (below): *Silo-capped studio*

9.21 (left): *Looking up from the interior at a compression ring for a yurt roof.*

heavy wood compression style roof, a continuous bond beam would be appropriate. If you want to use Velcro plates instead on a large diameter building, the roof members will need to be supported by an internal wall or post and beam structure that will direct the weight of the roof downward instead of outward — mimicking a shed-style roof (Fig. 9.18 & 9.19).

Dirtbag Silo

Another option for roofing a round earthbag building is to use a pre-manufactured metal silo roof. Metal silo roofs come in a multitude of sizes and can be purchased new (independently from the silo) (Fig. 9.20).

Dirtbag Yurts
(Lightweight Compression Roof System)

Yurt-style compression roofs are built with a compression ring at the apex of the roof and a tension ring through the eaves where they rest on the walls. The compression roofs for yurts are usually made of light two-by-four or two-by-six rafters. The tension ring is usually in the form of a cable run through the rafters on the inside of the walls (Fig. 9.21 - 9.23).

Some companies that manufacture yurts will provide all the components you need for just the roof. Commercial yurt roofs come in roof spans up to 30 feet (9 m) in diameter. Rafters are sized according to snow load calculations.

A yurt roof system can be adapted to earthbag walls by extending the rafters out to provide eaves — custom notching them to set flush on the edge of a Velcro plate and cinching them down with strapping (Fig. 9.22).

Choose any of the strapping methods described earlier in this chapter.

Yurt roofs can be insulated and they can be dismantled and transported. Using pre-manufactured roofing components has some advantages:

+ Precut-no waste on site.
+ Pre-engineered-a plus when building to meet code.
+ Easy to assemble-somebody else did most of the thinking.
+ Can be delivered to your site.
+ Usually a time saver.
+ Most come with some type of warranty.

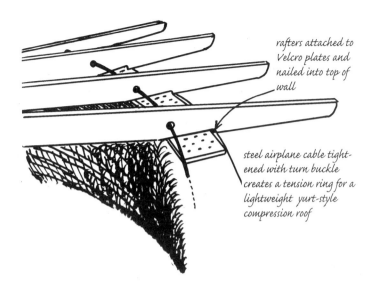

rafters attached to Velcro plates and nailed into top of wall

steel airplane cable tightened with turn buckle creates a tension ring for a lightweight yurt-style compression roof

9.22: *Light-wood, compression-style roof with extended eaves on earthbag walls.*

To us, the reciprocal roof is an ideal, exciting, and beautiful way to cap a round house without a bond beam. (Fig. 9.25).

For more information on reciprocal roofs, including engineering packages for self-building small round houses, homes, and community buildings, look in the Resource Guide in the back of this book.

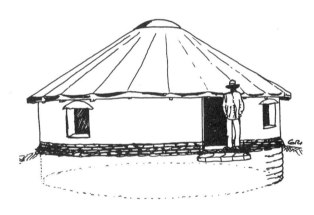

9.23: *Sunken earthbag round house with a commercially available yurt roof.*

The Reciprocal Roof

The reciprocal roof is a self-supporting spiral. An ingenious example of geometric harmony, it was re-introduced by a gentleman named Graham Brown. When pressure is applied from above, the spiral twists upon itself rather than spreading apart, the opposite dynamic to a compression roof (Fig. 9.24).

9.24: *Looking up into the eye of a reciprocal roof.*

9.25: *Inspired by the reciprocal roof, Jason Glick designed this fanciful gazebo lined with a sunken earthbag seating area for the Youth Garden Project in Moab, Utah.*

A Circle Within a Square

You may also put a square roof on a round building. When using dimensional lumber or pre-manufactured products, it may be the most efficient way to roof a round building. Place posts in a square around the exterior of the round walls. Connect them over the top of the walls with beams. Run two logs side-by-side across the center of the circle and add one log on the top of the first two. Strap down the logs. Extend rafters from on top of the center log out over the rectilinear beams (Fig. 9.26).

As an alternative, omit the center logs and run trusses across the whole width. Infill all the gaps in between rafters with a final row of bags and barbed wire. Sheathe however you prefer.

Multiple round spaces and free-form walls can also be covered using rectilinear roofing strategies (Fig. 9.27).

Adapt to suit your needs, taste, and climate. Let your imagination run wild!

9.26: *Low profile square roof supported on exterior post and beam framework with optional straw bale "living roof."*

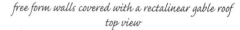

free form walls covered with a rectalinear gable roof top view

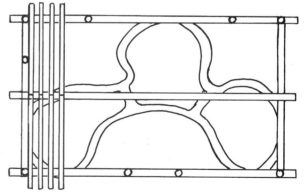

9.27: *Exterior post and beam structure provides interesting outdoor, protected living spaces.*

Roof Coverings

Use anything you want. Whatever we have on hand that is cheap, easy, recyclable, plentiful, enjoyable to work with, and low or non-toxic, is what we like best. Alison Kennedy collected leftover asphalt shingles, some free, some half-price out of stock colors. She is a knitter, so she arranged them in a playful pattern (see Fig. 9.1).

Roof coverings for earthbag buildings can be thatch, terracotta tile, cedar shingles or shakes, license plates, smashed cans, discarded metal roofing, or whatever you can imagine or scrounge. New products are being made from tires and wood chips to create Eco-shingles. Your own ideas are infinitely better than anything we might tell you.

A Living Roof (Fig. 9.28)

The use of living roofs has a long history that extends throughout almost every continent. North American Indians built a variety of buried pit houses protected by sod. Europe has a tradition of living roofs that come abloom with wild flowers in the spring. The benefits of a living roof are succinctly described by Christopher Williams in his book, *Craftsmen of Necessity.*

> As the seasons pass, the sod perpetuates itself; root intertwines root, and the roof becomes a solid whole which rain and weather only strengthen. In the winter the dead stalks of grass hold the snow for effective insulation. The spring rains beat the grasses down, so that they shed the excess water; then bring the roof to life again. The summer grasses grow long and effectively reflect the sun's heat.

A living roof is obviously very heavy. Any roof structure will need to be built accordingly to support it.

Another version of a living roof utilizes straw bales as a substitute for sod. By allowing the straw to compost over time, an ideal environment is developed for the propagation of indigenous grasses and flowers. A simple method for waterproofing a living roof is to cover the surface with a roll of EPDM type pond liner, or suitable substitute, followed by a layer of tight-fitting straw bales. After the bales are in place, clip the strings. This creates a beautiful, simple, single-layer roof with exterior insulation (Fig. 9.29).

9.28: *A living roof in the mountains of Colorado.*

WENDY EARTH-WATER PETERSON

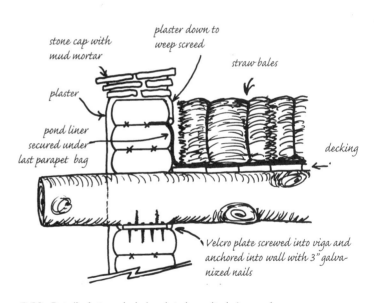

9.29: *Detail of straw bale insulated, vaulted viga roof.*

In a dry, windy climate, a straw-bale-insulated roof may need anchoring to keep it from blowing away. A simple method is shown here (Fig. 9.30).

Throughout the Southwestern United States, traditional adobe structures had their roofs protected by a thick layer of natural earth, supported by vigas

to reduce bulk, taper the bales that extend onto the eaves

vaulted viga roof without parapets

suspend rocks (or heavy chains) to sheep fence over bales as safe guard against wind

9.30: *Straw bale insulated roof kept in place under sheep fence, weighted down with suspended rocks.*

hefty 5" layer of adobe over parapet

note: pondliner installed under parapet bag

latillas provide exterior shade

poured clayrich adobe
pondliner
cardboard "cushion"
over 3/4" pumice
latillas

viga

viga screwed to velcro plate, plate velcroed into bag wall with 3" galvanized nails

9.31: *Exterior insulated, poured adobe roof for dry climates.*

and latillas. Any form of solid insulation that can withstand the weight of poured adobe can be used, such as rigid foam, straw bales, cans, bottles, scoria, pumice, etc. (Fig. 3.31).

Making Use of Local Resources

The roof reflects the climate more than any other part of a house. Here in the desert, raw earth can be our most resilient roofing material, while the Northwest coast offers an abundance of trees as an indigenous canopy for a roof built from timber. Use what you have where you are. Import as little as possible. Look at the way native people build things in the kind of climate you live in. Adapt. Let the environment be your esthetic guide (Fig. 9.32).

VIGAU HAMILTON

9.32: *With its thatched roof and post and beam with earth infill, the Baul House in Baulkutir, India, integrates naturally with its environment and the community it serves.*

CHAPTER 10

Arches: Putting the *Arc* Back into Architecture

Ahhhh … the wondrous arch! Every year millions of people travel thousands of miles to visit Arches National Park in Moab, Utah, just to stand in the presence of one of nature's most awe-inspiring sculptural forms: the arch. They're not coming here to see boxes or I-beams. They come to witness magic, the magical enduring beauty of the sweeping curve. Arch comes from the word *arc*, defined as "the part of a circle that is the apparent path of a heavenly body above and below the horizon." Like a rainbow! It is no accident that the most sacred places of worship incorporate the curve.

10.1: *Delicate arch in Arches National Park, Utah. This arch measures 33 feet (9.9 m) between the legs, and is 45 feet (13.5 m) high.*

There is something special about walking through the curve of a vaulted doorway or gazing out of a curved window frame. Standing under a domed roof one often experiences a sense of boundlessness. In a round house you will never feel cornered.

We live in a land of ancient arches sculpted by the forces of nature. Yes, we consider nature the ultimate artist. Humans are often inspired by nature's artistic ingenuity. Nature is both artist and engineer. Thousands of years ago people discovered that the beauty of an arch also lies in its structural integrity. You might call it sculpture with a purpose. An arch is amazingly strong. When people began using the arch in their structures, a new definition for the art of building was born. We do not call it box-itecture or linea-tecture. It is the arch that inspired the birth of *Architecture*.

The Dynamics of an Arch

Our love affair with earthen architecture began with building our first arch. The excitement mounts as we tamp the keystones into place, anticipating the removal of the forms to reveal the magic space within. What is it that gives the arch its magical qualities? The forces

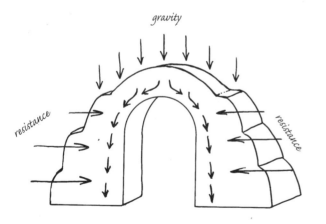

when the forces of compression from overhead are met with the forces of resistance from either side, the resultant forces are transferred towards the ground rather than out to the sides

10.2: *The dynamics of an arch.*

acting upon it and how they relate to one another define the apparent gravity defying nature of the arch.

According to Webster's New World Dictionary, the meaning of the word dynamic comes from the Greek *dynamicos* meaning power, strength. It is further defined as "the branch of mechanics dealing with the motions of material bodies under the action of given forces." The definition of arch is "a curved structure that supports the weight of material over an open space, as in a doorway, bridge, etc." An arch is held in place by two opposing forces. The force of gravity pulls downward (*compression*), while the force of *resistance* from either side prevents gravity from flattening the arch. Resistance is a force that retards, hinders, or opposes motion (Fig. 10.2).

All arches that are built by stacking units (adobe, brick, stone, earthbag, etc.), are utilizing compression and tension as dynamically opposing forces designed to provide structural integrity.

This gravity defying force on either side of an arch is referred to as *buttressing of an arch*. An arch relies on sufficient buttressing to maintain its shape. As the force of gravity pushes down on the weight above the *keystone* at the top, the resultant force is transferred to the sides, where it is met with the resistance of the buttress or adjacent walls that provide the same resistance. The tension created by these two opposing natural forces has given us some of the most stunning architectural features created by man or nature (Fig. 10.3).

The nature of the earth as resistance coupled with the eternal pull of the earth's gravity is combined in arch and dome construction. This is the magical dynamic of an arch. The very force that pushes and pulls arches, domes, and vaults down also holds them up. Gravity is a structural component of this architecture.

Two Classical Arches
The Roman Arch (Fig. 10.4)

The hemispherical arch has been in use for over 6,000 years. It is generally referred to as the *Roman* arch, as it was during the Roman Empire that it was used extensively for bridge building. The skilled Etruscan engineers taught the Romans the use of the keystone arch, enabling them to build extremely strong and durable bridges. The idea is quite simple.

Imagine a ring of tapered stone blocks arranged in a circle. If one were to take a rope, wrap it around the ring, and tighten it, all it would do is force the stones more tightly together. Exchange the rope for steel cable and tighten by twisting with a steel bar shoved between the cable and the ring and the circle of stones just becomes stronger! A mighty force would have to be used to destroy a ring constructed like this.

To create a keystone arch, one half of this ring of stones was simply stood up on its end. In a typical Roman arch bridge, these ends rest on piers made of stone blocks mortared together with pozzolanic cement. Sufficient buttressing or adjoining walls provide the tension, as the rope or cable did for the ring of stone. The weight of the stone and the bridge itself compress the tapered stones together, making the arch an extremely strong structure.

10.3: *Landscape Arch in Arches National Park, Utah. This arch measures 306 feet (91.8 m) across and is 106 feet (31.8 m) high.*

10.4: *Built with cut sandstone, this Roman arch (or voussoir) in Moab, Utah, celebrates its 100th birthday.*

Heavy wagons and legions of troops could safely cross a bridge constructed of arches without collapsing the structure. Many of these bridges outlasted the Roman Empire, the Dark Ages, Middle Ages, and on into modern times, serving General George Patton during World War II just as they had served Caesar almost two millennia before (Fig. 10.5).

10.5: *Built during the Roman Empire, Pont du Gard typifies the use of the keystone arch in bridge building. The lower span is still used as a roadway, while the upper arches functioned as an aqueduct for several centuries.*

  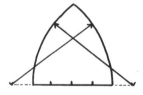

half circle or
Roman arch

8-point Gothic or
Egyptian arch

lancet arch

10.6: *The placement of the compass point determines where the springline begins and defines the shape of the arch. The further the compass point is placed from the center of the baseline, the taller and steeper the profile becomes.*

The Gothic Arch

The second type of classical arch we will address is called the *Egyptian* or, more recently, *Gothic* arch. In the 1200s, Abbot Suger, of the Abbey of St. Denis outside of Paris, had a plan for transforming the squat, heavy Romanesque style into an architectural wonder of the time. This is a steeper-sided arch than the hemispherical shape of the Roman arch. While the same forces of compression and tension are at work on both arches, we will soon see how the steepness of the Gothic arch directs the forces of compression at a steeper angle, and how that affects its performance compared to a Roman arch.

There are a number of different types of arches, whose names are mostly self-explanatory: elliptical, flat, horseshoe, lancet, obtuse, ogee, segmental, semicircular, etc. All of these arches share the same principles of geometry but, for simplicity's sake, we will address the two classical arches, Roman and Gothic. These are the types of arches we have successfully worked with since we made our first dry stack arch. That's not to say the other arch forms would not work with earthbag construction; we just haven't yet tried any others!

The First Step: Drawing the Arch (Fig. 10.6)

To make a Roman arch (on paper), draw a straight line the width of the desired arch. Place the pivot point of an architectural compass at the center point of this line. Extend the pencil end of the compass to either end of the drawn line. Sweep the pencil around to the other end of the line. A Roman arch is a perfect half-circle.

A Gothic arch has a much steeper pitch than a Roman arch. For example, an eight-point arch is one type of Gothic-style arch. To calculate an eight-point arch, divide the width of the base into eight equal segments. Place the stationary end of an architectural compass on the first point in from one end of the base. Extend the pencil to the opposite end of the base. Scribe an arc to complete a half-curve. Repeat this same procedure from the other side of the base.

Where the two arcs intersect above the center of the arch denotes the shape of this arch. The same principle of the eight-point arch can be used to make many different arch configurations. Just change the pivot point of the compass to other locations along the base, even outside the area defined by the arch. This pivot point denotes the angle of the springline of an arch.

Springline of an Arch

The inward curve of an arch begins at the *springline*. In earthbag construction, this is where we shape our first *fan bag* to initiate the beginning of an arch. To demonstrate where the arc of the springline begins on a Roman arch form, attach a string to the center bottom edge of the half-circle. As the string is held to the outside edge of the arch, it shows the angle at which to contour the bags around the form (Fig. 10.7).

Because Egyptian arches are considerably steeper-sided than Roman archs, they only need a small amount of extra wedge shape to get them started following the shape of the arch. After the first fan bag sets the initial pattern, the consecutive bags will be almost rectilinear in shape until they reach the *keystone* slot at the top of the arch form (Fig. 10.8).

Determining the Outward Forces of an Arch

The steeper the sides of an arch are, the stronger it becomes. A steep-sided arch transfers the weight above it at a more vertical pitch than a shallow arch does. The shallower the pitch of an arch, the more pressure it forces to the sides, creating horizontal stress. Therefore, the shallower the arch, the more it needs to be buttressed to counteract this horizontal push.

Here is a simple way to determine the amount of buttressing necessary for a given arch shape (Fig. 10.9).

The distance between points B and C is equal to the distance between points C and D at only one point on any given arch shape. Placing the pivot point of an architectural compass on point D and the scribing end on point C, rotate the compass in a 180° arc. Where the circle intersects a straight line drawn between points C and E designates the amount of buttressing necessary for a given arch shape (Fig. 10.10a & b).

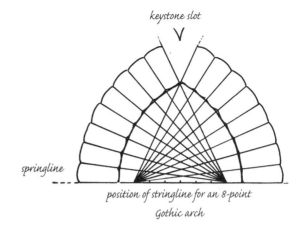

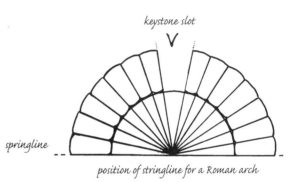

10.7: *Examples of a Gothic and Roman arch with stringlines and keystone slots.*

10.8: *The attached stringline assists as a visual aid in determining the appropriate angle to shape the fan bags.*

It's easy to see how much less buttressing is necessary for a more steeply pitched arch. The same holds true for a dome, which is an arch rotated 360°.

In a straight wall, we need to allow ample distance between any openings to ensure enough solid walls to act as buttressing. In a round wall, we could safely build a string of arches one after another, as in the Roman Coliseum, as the horizontal force of each arch is counteracted by the horizontal force of the next, and so on and so forth in a perpetual wheel of support.

The Catenary

A curve described by a uniform flexible chain hanging under the influence of gravity is called a *catenary*. In Medieval times, masons used their squares and com-

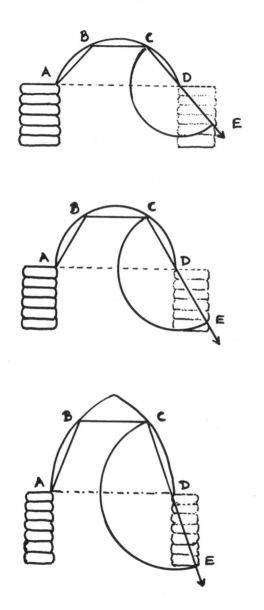

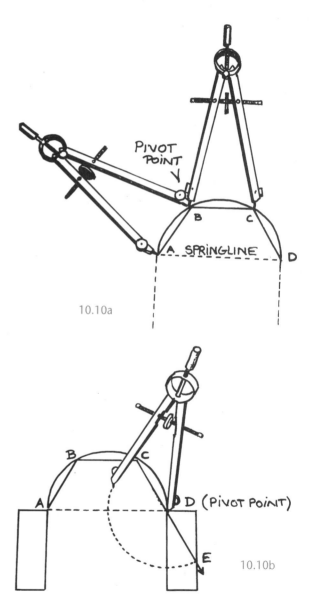

10.9: *Point E (where the arrow exits the half-circle) determines how thick the wall buttressing needs to be to support this shape of arch.*

10.10a & b: *After using the compass to locate the equidistant points of A, B, C, and D, create the arc between points C and E, using point D as the pivot point.*

10.11a (above): *A chain hung between two points creates a catenary curve, an arc in complete tension.*

10.11b (right): *When the catenary curve, in Figure 10.11a, is turned upside down, it creates an arc in complete compression.*

passes, and probably chains, to create geometric shapes in stone. They didn't study theorems and proofs, but instead found natural shapes that stayed in balance.

In 1675, the English scientist Robert Hooke determined how those early masons accomplished seemingly magical feats of masonry. Basically, Hooke said that if you hang a chain from two points, it naturally hangs in complete tension with zero compression. That is to say, the tension between each link is the same for every link. Now, if we were to fuse each link together and turn the

shape of the hanging chain upside down, we would get the shape of an arch in complete compression (Fig. 10.11a & b).

Hang a chain over the drawing of an upside down arch and you will see where the outward forces of thrust are. **The chain must hang within the center one-third of the arch and supporting walls in order to cancel out any bending forces that want to push the walls outward.** In arches with shallow profiles, the horizontal thrust tends to force the legs apart. That is why buttressing is vital to arch building, and arches built with earthbags are no exception (Fig. 10.12a & b).

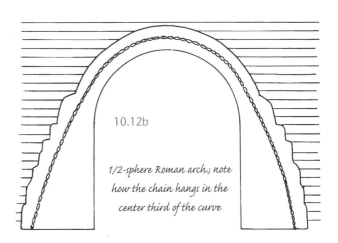

10.12a

interestingly, the hanging chain correlates with the compass formula used for determining the amount of buttressing needed for this 8-point arch

10.12b

1/2-sphere Roman arch; note how the chain hangs in the center third of the curve

10.12a & b: *Within the lines of force of every successful arch resides the imprint of the catenary curve. Every shape arch, whether it is tall and slim or shallow and thick-walled, is designed to contain within its core the natural equilibrium of the hanging chain.*

This is another way to show that the lower the profile of an arch, the more the buttressing. A more steeply pitched arch requires less buttressing; the horizontal thrust is countered by the tension of the buttressing. Once an arch (and its buttressing) accommodates the catenary curve, stress analysis is achieved without using a mathematical formula. To build arches of stacking units, whether adobe, stone, or earthbag, you begin by *seeing* them, not calculating them. Arched openings built with stacked units are concieved in a very different area of the human psyche than are steel, concrete, and wood rectilinear openings. That is probably why they *touch* a different part of the human psyche, as well.

Vaults

A *vault* is essentially a really deep arch, like a tunnel. There are two strategies for building vaults. *Keystone vaults* use the same form work we use for supporting arches until the keystones are installed. The forms must support the full length of the vault. Because of forces directed outwards from the keystones, this style of vault requires a tremendous amount of buttressing.

This is the type of vault we built into the entrance of the Honey House.

To create a vault, we extended our box and arch forms out to accommodate the extra length, while, at the same time, we fortified the width of the walls with buttressing to counteract the compressive forces caused

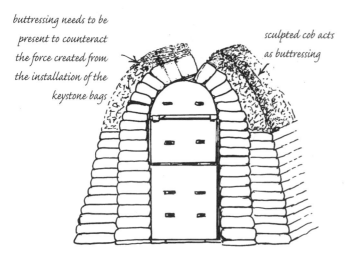

buttressing needs to be present to counteract the force created from the installation of the keystone bags

sculpted cob acts as buttressing

10.13: *Without additional buttressing on either side of the vault, the keystones would have forced the walls apart beneath the arch.*

10.14: *Nubian masons incorporated the catenary curve in the construction of leaning vaults.*

by the installation of the keystones (Fig. 10.13). A dormered arch in a Gothic shape would direct the forces of compression downwards through the legs rather than out towards the sides, requiring less buttressing than a Roman arch.

Leaning Vaults

The other style of vault construction is called a *leaning vault*. It was developed by Nubian builders as a way to build vaults with less material and zero formwork. A leaning vault transfers most of its compressive forces to whatever it is leaning on at either end rather than out to the sides, like a keystone vault. The wall that the vault leans on is its buttressing, and must be of substantial thickness to counteract the weight of the leaning vault. A leaning vault can be built up against a thick, vertically plumb wall, or the inclined wall of a dome (Fig. 10.14).

Free-standing leaning vaults are easier to build with bricks than earthbags. Small earthbag leaning vaults can be built to dormer the entrance into a dome, however, it is difficult to build a leaning vault out to a plumb (vertical) wall surface with earthbags alone.

We have not experimented with free-standing vaults of any kind, which is the main reason we don't present them with any depth in this book. We find the idea of building leaning earthbag vaults arduous. The inclined position of the bags or tubes makes tamping them at an angle awkward. Scaffolding would be required as the slope would be too steep to stand on, installing barbed wire would become a juggling act. In general, it would not meet FQSS principles. Other than as short dormers, it would be easier to build earthen vaults the way the Nubian masons have done for centuries, with adobe brick.

This freshly laid Earthbag wall was constructed in a weekend at the 1999 Colorado Natural Building Workshop held in Rico, Colorado (Fig.1, left).

Serena Supplee and Tom Wesson's Earthbag walls built on a busy street corner provide privacy, sound protection, as well as a built-in flower planter. Moab, Utah (Fig. 2, below).

The Honey House with exposed bag work and temporary wooden arch forms is a self-supported corbelled earthbag dome (Fig. 3, below left).

The two feet sunken interior serves as a cozy drafting studio (Fig. 4, right).

The exterior is protected with a red clay plastered roof and lime plasters on the vertical wall surfaces. The sculpted rain gutters direct water down the buttress away from the foundation (Fig. 5, below).

Earth**h**b**a**g walls make attractive sturdy privacy walls, sound, and wind barriers that can be adapted to suit many climates. Mitchell **May**'s 300 yard long earthbag wall is covered with a cement/lime **p**laster in Castle Valley, Utah (Figs. 12, top and 13, right).

Carol Owen's serpentine wall with arched windows sports a hand-**ru**b**b**ed lime 'sandstone' finish, Moab, Utah (Fig. 14, above).

This officially permitted earthbag home was tastefully designed and built on a shoestring budget by owner/builder Alison Kennedy. A conventional hip style truss roof protects the lime frescoed wall plaster (Fig. 6, above).

The interior highlights an open kitchen, poured adobe floors, and earth plastered walls throughout (Fig. 7, left).

This gracious home integrates earthbag tube walls and shallow 'boveda' brick domes hidden by parapets designed and built by Mara Cranic in Baja Sur, Mexico (Fig. 8, right and Fig. 9, below center). The sprayed on cement stabilized plaster accentuates the contours of the tube walls (Fig. 10, below left).

Susie Harrington and Kalen *Jones* staggered multiple tiers of *earth-bag* retaining walls to provide level gardening space with *grace*. Moab, Utah (Fig. 11, below *right*).

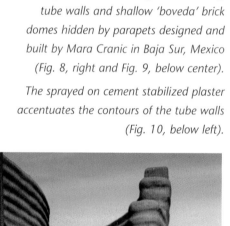

Sarah Martin and Monty Risenhoover constructed a 65-feet by 20 feet living space onto the back of the Comb Ridge Trading Post in Bluff, Utah. The interior features rich, Chinle red plaster, ponderosa pine vigas, and a fired brick inlayed floor (Fig. 15, top left)

The interior of this 750 square feet Ranger Station built by the Bureau of Land Management outside Bluff, Utah is finished with colorful, wild-harvested earth plasters collected from the public lands nearby (Figs. 16 and 17, left and above right).

The buttressed and bermed earthbag wall provides structural stability and support for a drying rack in this earthbag/strawbale hybrid greenhouse built by the Youth Garden Project - Moab, Utah (Fig. 18, right).

Eddie Snyder's earthbag flower planters adorn the entrance to a popular restaurant in Moab, Utah (Fig. 19, below left).

Let the truth be known – an earthbag truth window reveals all! (Final image, below right).

CHAPTER 11

Dynamics of a Dome

"Straight is the line of duty, and curved is the path of beauty." — Hassan Fathy (1900-1989)

This is, we feel, where earthbags exhibit their greatest potential; to us, it is the essence of earthbag building. We are able to build an entire house from foundation to walls to roof using one system. (To gain greater understanding of the dynamics of a dome, please read "The Dynamics of an Arch," in Chapter 10, to better acquaint you with some of the same language and principles that are inherent in both.

We have talked about and demonstrated the earthbag wall building system. We have extolled the virtues of turning corners into curves and the magic of the circle versus the square. Now we will put it all together to build one of nature's most sophisticated engineering achievements, the dome.

11.1: *The mosque at Dar al-Islam in Abiquiu, New Mexico, designed by Hassan Fathy.*

Dynamics

Nature is the ultimate utilitarian. She combines function with form, using the simplest strategy to get the highest level of structural integrity with the least amount of materials. Take a fresh raw egg, place the ends in both palms and squeeze with all your might. This thin, seemingly fragile membrane will resist your effort. Like an egg, a dome is designed using a double curvature wall; it curves in both the horizontal and vertical plane at the same time.

To better understand the dynamics of a double curvature wall, let's compare a dome with a cylinder and a cone.

A cylinder (a vertical round wall) curves in one direction, horizontally, while the sides remain vertically plumb. A cone also curves in one horizontal direction and, although decreasing in circumference at one end, it still maintains a linear profile. The dome curves both horizontally and vertically, producing a spherical shape. Technically, a dome's profile can be many shapes, varying from a low sphere to a parabolic shape, like the opposite ends of an egg (Fig 11.2).

Earthen domes rely on two opposing natural forces to hold them together: gravity and tension.

This balancing act of downward pressure meeting perimeter resistance is a very sophisticated engineering technique and has been employed for millennia. Domes built from individual units such as

adobe block, stone, and, in our case, earthbags (or tubes) use gravity and resistance as integral structural devices. These forces differentiate them from a geodesic or cast concrete dome that rely on a monolithic framework to hold them together (Fig. 11.3).

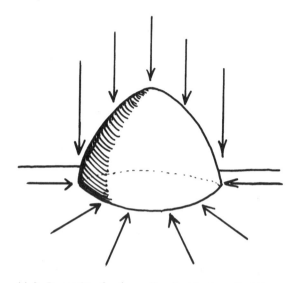

11.3: *Dynamics of a dome: Gravity tries to pull all the mass of the earth overhead down (compression), while the outer perimeter of the dome rests on a "ring of tension" that resists the spreading of the walls.*

Like an arch, a dome is only as strong as its buttressing. A dome, however, is self-supporting and does not need any structural formwork other than the arched window and door openings. A dome is an arch built in the round. The point where the springline begins is where buttressing needs to be present.

Traditional Dome Building Techniques

Corbelling is a technique established thousands of years ago. It involves laying the courses of brick horizontally rather than inclined. Each course is stepped in slightly from the preceding course, actually cantilevering over the underlying bricks. This creates a profile more cone-like in shape rather than hemispherical (Fig. 11.4).

Another traditional dome building technique using adobe bricks is called *Nubian* style masonry, a

single curvature cylinder wall

single curvature cone wall

double curvature dome wall

11.2: *Comparison of three shapes: A cylinder, a cone, and a dome.*

technique developed by builders in Upper Egypt over three thousand years ago. They installed adobe bricks at an angle, beginning at the springline, which enabled them to follow a low profile hemispherical shape. As the bricks follow the profile, the angle of the bricks is almost vertical when they reach the top of the dome (Fig. 11.5).

The more common designs, used for adobe brick domes in the Middle East and the Mediterranean, have vertical walls that are either round or square and come up to head height from grade level, at which point a shallow, hemispherical, or parabolic shaped dome is constructed.

In traditional techniques, the horizontal thrust of a shallow dome roof is counteracted by increasing the thickness of the walls two or three times the thickness of the dome, providing extra mass as buttressing. Building multiple domes that share the same wall cancels out the horizontal thrust where the domes meet. Lateral buttressing can be strategically incorporated for extra stability (Fig. 11.6).

Bond Beams

Today, a hemispherical, low-profile brick dome uses a concrete bond beam at the springline as a *tension ring* to counteract the pressure caused by the horizontal forces created. The lower the profile of the dome, the more pressure is exerted out to the sides. Without the continuous tension ring created by the bond beam, the horizontal thrust from the dome would push the walls out from under it, and collapse (Fig. 11.7).

Mexico has a traditional style of building brick dome roofs called Bovedas. Boveda domes are similar in construction to Nubian masonry, where the bricks are set at a progressively shallower angle as the dome closes in at the top (Fig. 11.8).

11.6: Traditional techniques for counteracting the horizontal thrust of a shallow, dome roof; increase the thickness of the walls, build multiple domes, incorporate lateral buttressing.

11.4: These corbelled stone and brick domes, built in the 15th century in Zacatecas, Mexico, are still used today by the local villagers to store grain, and for other purposes.

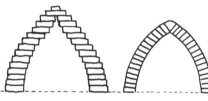

corbelled dome stacked-units are stepped in gradually, creating a tall pointed shape dome

Nubian style units are mortared at an increasingly steeper angle to form this Egyptian shape dome

Nubian style units are inclined at an extreme angle to form this Roman half-sphere dome

11.5: Corbelling compared to Nubian style dome building techniques

We are explaining terms and techniques used in more traditional and contemporary earthen dome building to acquaint you with the language, and to point out the differences (and similarities) as compared to an earthbag dome.

Earthbag Domes

The construction of Nubian style adobe brick domes is ingenious, but extremely difficult to replicate with earthbags. We are limited to the *corbelling* technique, based on the nature of working with the bags themselves. By corbelling the rows of bags or tubes, we maintain a flat surface to stand on while filling and tamping the bags in place — meeting FQSS principles. As the bags or tubes are stepped in every row a little at a time (gradually decreasing the diameter), the walls eventually meet overhead to form the roof.

Earthbag domes need to be steep-sided as the rows can only be stepped in a specific amount that is determined by the size of the bags (or tubes) used. Corbelling an earthbag dome results in its characteristic shape (as tall as it is wide), resulting in an essentially *lancet* or parabolic style arch in the round (Fig. 11.9).

11.7 (above): *The concrete bond beam acts as a continuous ring that inhibits the walls of the dome from being forced out.*
11.8 (below): *Bovedas are built at a seemingly impossible low profile, with the bricks often left exposed on the inside.*
Bovedero: Ramon Castillo. Credit: Mara Cranic

11.9: *The Honey House, a corbelled earthbag dome in Moab, Utah.*

All of the design strategies used for typical brick dome building are simplified when designing an earthbag dome. The simplest way to provide the most solid form of buttressing is to begin the *springline* at or below grade level. The springline begins at the point where the first row of bags is stepped in. By beginning the springline close to the ground, we omit the need for a concrete bond beam as the surrounding earth provides us with a natural tension ring sufficient to buttress the perimeter of the dome.

We actually do two things to aid stability: lower the springline, and steepen the profile. By raising the profile to a steep angle, the horizontal forces are minimized. The steepness of the profile helps to direct the gravitational forces down towards the ground rather than out to the sides.

As a rule of thumb, earthbag domes are designed with a compass formula that produces a shape as tall as it is wide. For example, a 20-foot (6 m) interior diameter dome will also be 20 feet (6 m) high at its peak, from springline to ceiling. We have chosen a simple formula using an architectural compass that creates a subtle curve with a steeply pitched profile. For the sake of security coupled with the limitations of the earthbags themselves, we like to use this formula for designing a dome.

Designing (Drafting) an Earthbag Dome

How to Use an Architectural Compass

An architectural compass is the dome builder's friend, so let's get acquainted with the new toys we will need to design an earthbag dome on paper. We will need:

- An architectural-student-quality drawing compass (preferably with an expandable arm)
- A three-sided architect's (or engineer's) scale ruler (in inch or centimeter increments)
- A good mechanical pencil and eraser
- A two-foot (0.6 m) long T-square
- A combo circle template (optional)
- A flat surface with a square edge like a pane of glass or Plexiglas or a sheet of Masonite

(or drafting board) to use as a square edge for the T-square to follow
- Some light-stick tape for tacking down our drawing paper without tearing it (Fig. 11.10).

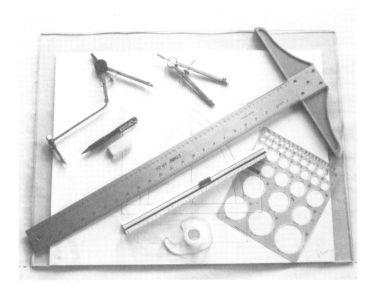

11.10: *Drafting tools for desiging an earthbag dome.*

We also like to have a sheet of graph paper taped to the surface of our plate glass drafting board to use as a quick reference for aligning drawing paper.

As a rule of thumb, all measurements for domes (and circular structures) are defined by their interior diameter. This measurement remains constant, whereas the exterior measurements may vary according to what thickness the walls become. There are many bags and tubes with different size widths available, and many different ways to finish the exterior of an earthbag dome. For practice, let's design a small 14-foot (4.2 m) interior diameter dome using a half-inch (1.25 cm) scale.

Align your drawing paper with the center of the drawing board and tape the corners. Using the T-square as a guide, set the ruled architect's (engineer's) scale against it horizontally on the page and mark the centerline on the paper where you want the ground level of your structure to begin. Leave a few inches of free space at the bottom of the paper. For now, we are going to design a dome with a floor level that begins at grade level. Using the scale of one-half-inch (1.25 cm) equal to one-foot (30 cm) ratio, draw a line equal to 14 feet (4.2 m), with seven feet (2.1 m) on either side of your center mark. This is the interior diameter of the dome (Fig. 11.11).

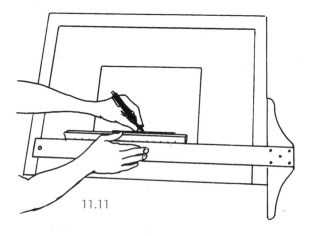

11.11

Turn the T-square vertically on the board and place the half-inch (1.25 cm) scale ruler along the center mark. From the center of the diameter, measure 14 feet (4.2 m) up and draw a line. This line denotes both the interior (vertical) height of the dome (14 feet [4.2 m]) and the (horizontal) radius (7 feet [2.1 m]) (Fig. 11.12).

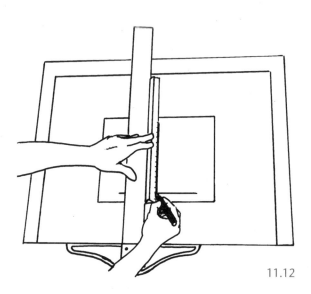

11.12

We are going to mark the spot where our compass pivot point will be positioned. The compass formula is one-half of the radii from the interior diameter width of the dome. Divide the radius in half (7 feet [2.1 m] divided by 2 = 3.5 feet [1.05 m]). Make a mark three and one-half feet (1.05 m) beyond the end of your interior diameter line. Repeat this process on the opposite side of the horizontal line (Fig. 11.13).

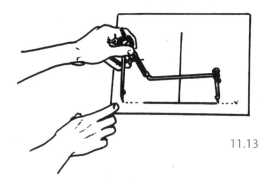

11.13

Use the architectural compass to create the outline of the dome. Adjust the spread of the compass so it will reach from the fixed pivot point (3.5 feet [1.05 m] beyond the end of your horizontal line) to the other end of the 14-foot (4.2 m) diameter horizontal line. This is where our springline will begin. With steady pressure, swing the compass arm up to meet the 14-foot (4.2 m) height mark (Fig. 11.14a).

Reverse the process to complete the other side (Fig. 11.14b).

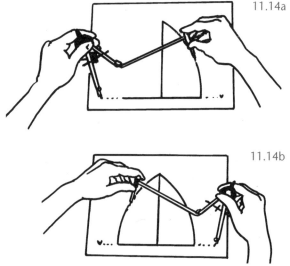

11.14a

11.14b

Adjust the compass to accommodate the thickness of the bag walls you will be building. As an example, for a 50-lb. bag with a 15-inch (38 cm) working width, use the scale ruler to extend the 14-foot (4.2 m) diameter base line 15 inches (38 cm) more on either side. From the original fixed compass point, lengthen the arm to reach to the exterior wall mark on the opposite side of the dome and swing the free end of the compass arm up to the top of the dome (Fig. 11.15).

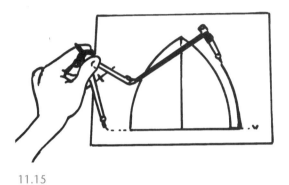

11.15

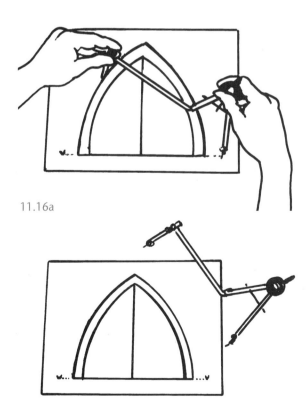

11.16a

11.16b

Repeat for the other side of the dome (Fig. 11.16a).

We now have the basic shape of the dome with its interior, exterior, width, and height measurements (Fig. 11.16b).

We'll do a drawing of the same structure highlighting many different features, always beginning with the same basic shape we just drew. We'll do an elevation drawing that shows what the building looks like from grade perspective, and another that highlights a sunken floor or underground room. Another drawing will show window and door placements, or roof details, etc. But first, let's establish the measurements we'll need to transfer onto the construction-size building compass to create the same profile we drew on paper. (See Chapter 3, page 48).

Ideally, this drawing is easiest to read when done on a large piece of paper or cardstock using a large scale, like one inch (2.5 cm) equals one foot (30 cm). Or work in metric, if this is easier for you. If you don't have a drawing compass large enough to draw this expanded profile, then you can scribe the profile with a pencil tied to a string (Fig. 11.17).

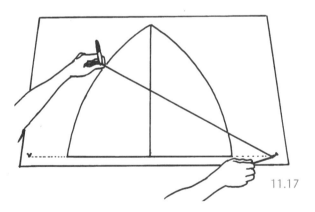

11.17

The centerline from top to bottom of the dome represents the construction-size building compass. Starting at the bottom, use the T-square and mark along its length every one-half-inch (which is the equivalent of six inches on our scale) of height or, if using centimeters, every 1.25 centimeters (which would be the equivalent of 10 cm).

Align the T-square horizontally along the first height mark. Draw a line from this height mark to the interior edge of the dome wall (Fig. 11.18).

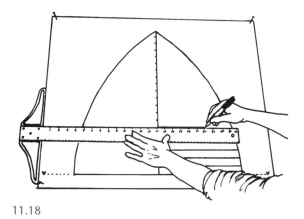

11.18

Measure this horizontal distance between centerline and interior edge of wall and write this measurement along the wall of the dome.

Round off the number to the nearest one-half-inch or centimeter. Repeat this process, going up the vertical compass line until each height mark has a corresponding radius measurement (Fig.11.19).

11.19

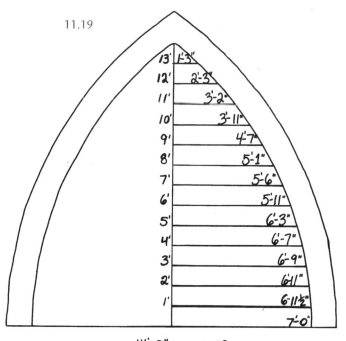

13'	1'-3"
12'	2'-3"
11'	3'-2"
10'	3'-11"
9'	4'-7"
8'	5'-1"
7'	5'-6"
6'	5'-11"
5'	6'-3"
4'	6'-7"
3'	6'-9"
2'	6'-11"
1'	6'-11½"
	7'-0"

14'-0" DIAMETER

These measurements are what you will refer to when adjusting the length (radius) of the building compass arm during construction. This process aids us in duplicating the profile from paper to reality.

Remember the Catenary arch? To test the dynamic forces of our compass formula we hang a chain on a cut-away view of our drawing. Check to see that the chain hangs well within the middle third of the wall. If the chain strays from the center (either inside or out) we can do two things: increase the thickness of the walls and/or free hand a new profile that more closely follows the shape of the hanging chain (Fig. 11.20).

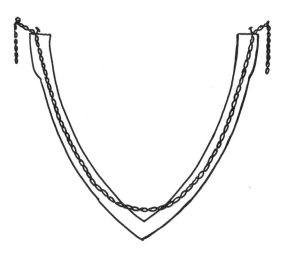

11.20: *The suspended chain should hang within the center third of the roof and walls, and, where the chain exits the compass profile, is where buttressing is added.*

You may be wondering why we don't calculate the compass measurements based on the height of each row of bags. At first, that is what we did (on paper). We measured the height of our working bags at five inches (12.5 cm), and proceeded to calculate all of our radius adjustments at five-inch (12.5 cm) increments. This worked fine at first while the lower bags were only stepped in one to two inches (2.5-5 cm). When we started overhanging them three to four inches (7.5-10 cm), the portion of the bag that remained on the underlying row started flattening down to four inches (10 cm) and eventually three and one-half inches (8.75 cm) in thickness (Fig. 11.21).

This is one of those phenomena that one only discovers by doing it. We couldn't count on the row of bags conforming to our role model on paper. Making pre-calculations provides reference points to help keep us on track along the way. We know, for example, that at the five-foot (1.5 m) height we want to be at a radius of six feet, three inches (187.5 cm) exactly, for this 14-foot (4.2 m) diameter dome. If a row finishes out at a height in between the marks on the compass, we just split the difference. Pre-measuring on paper gives us a reference guide that helps to speed the process during construction.

As long as we are within one-half-inch (1.25 cm) of our radius, we still feel fairly accurate. We are dealing with a mushy medium, after all, that will squish out here and there. It's partly the nature of the material. For the sake of creative license, and given unforeseeable circumstances, feel free to make alterations, as long as they maintain the structural integrity of the dome.

If all of this sounds terribly confusing and complex, it is simply because it has yet to become familiar to you. Liken it to trying to explain how to drive using a stick shift to someone who has never even sat in a car. After a while it will become automatic. Relax; you'll get it.

Advantages of Earthbag Domes

Structurally, the distinctive difference between earthbag domes and brick domes is their higher tensile strength, derived from the installation of two strands of barbed wire per row. In essence, the added tensile strength combined with the woven polypropylene fabric helps unify the individual rows into a series of stacked rings. Each of these complete rings creates a mild tension-ring effect, offering tension under compression throughout the whole dome not just at a single bond beam. Excellent results have been obtained from both load bearing and lateral exertion tests conducted in Hesperia, California (see Tom Harp and John Regner,

11.21: *The further the bags (tubes) are cantilevered over the underlying bags, the more they flatten out.*
Photo credit: Keith G.

"Sandbag/Superadobe/Superblock: A Code Official Perspective," *Building Standards* 62, no. 5 (1998) pg. 28). In Chapter 18 we give a condesed reprint of this article.

Functionally, the steeper the pitch is, the more quickly it sheds water. This is an added benefit in dry climates that get short bursts of violent storms. Rain has very little time to soak into an earthen clay or lime plastered roof covering.

Design-wise, the taller interior allows ample room for a second story or loft space to be included (or for a trapeze or trampoline for more athletic folk). A built-in loft can also double as a scaffold during construction. A dome also provides the most living area for the least surface area, compared to a comparably sized rectilinear structure.

Spiritually, the compass formula we use creates an arc within a perfect square. This is our personal opinion, but we have a strong feeling that the specific ratio of this shape acts as both a grounding device and an amplifier for whatever intentions are radiated within the structure. So pay attention to your thoughts.

Energy Consumption: as noted in Appendix D, a circle provides more area than a rectangle built with the same length of perimeter wall. A dome takes another step by increasing the efficiency of the ratio of interior cubic feet of space to exterior wall surface. Domes provide the greatest overall volume of interior space with the least amount of wall surface. A dome's smaller surface area to internal space ratio requires less energy to heat and to cool it. And because of the shape, air is able to flow unimpeded without ever getting stuck in a corner.

Climatically, earthen walls are natural indoor regulators. Earthen walls breathe. They also absorb interior moisture and allow it to escape through the walls to the outside, while at the same time help to regulate interior humidity. A hot, dry climate gains the benefit of walls that are capable of releasing moisturized air back into the living space. The same situation can benefit the dry interior created by wood burning stoves in a cold winter climate. Earth is a natural deodorizer and purifier of toxins. An earthen dome literally surrounds one with the benefits inherent in natural earth.

11.22: *Experiment with small-scale projects before tackling a large one.*

Disadvantages (depending on your point of view) of Earthbag Domes

If a large rambling structure is of great importance to you, an earthbag dome might not be the best choice. Just like in nature, every building has an optimum size for its own equilibrium. We feel that earthbag domes are best suited to small to moderately sized structures, up to 20 feet (6 m) in diameter. If you think about it, the wider the diameter the taller the roof, and the further you have to heave dirt up onto the wall.

Rather than building a single dome bigger, we'd rather build several, more manageably sized interconnected domes. Another way to reduce the height of a larger dome is to build one-third to one-half of it underground, or berm it into a hillside. If a big space is what really trips your trigger, we recommend building a kiva-style, round structure with vertical walls and a more conventional roof rather than an earthbag dome.

Another drawback to building domes is that there is no mention of earthen dome building in current building codes. If you live in an area where building codes are strictly enforced, you may have trouble getting a permit, unless you are able to get a licensed engineer to sign off on your plans. Regrettably, most engineers trained in North America have little or no experience building domes, even though dome building has been successfully pursued in other countries for centuries. With a little ingenuity and a strong commitment to what you want, there are ways to circumvent or bend the vaguely rigid stipulations presented in most building codes. (See Chapter 18 for more on this subject).

Building an earthbag dome is an arduous task that tests your ingenuity and resolve. Attempting a dome project by yourself, while possible if you build small, is difficult and sometimes frustrating without someone there to help you with the tough stuff. Earthbag domes are easier to build with a good size crew of five (or more) dedicated people, especially as the height of the building increases.

The bags or tubes need to be protected from the sun as soon as possible, as they are a critical structural component of this style of architecture. In very rainy climates, the walls may need to be protected from excessive moisture during construction to avoid a meltdown. All of these disadvantages are addressed in various parts of this book. We will go over more specific details in the next chapter. Sometimes, what we consider a disadvantage is our own misunderstanding of a principle, or fear of making a mistake that can't be rectified. What we've come to discover instead is that action dispels doubt.

There is no substitute for experience. We recommend taking a hands-on workshop before tackling a large-scale dome-building project. By all means, experiment with a small six- or eight-foot (1.8 or 2.4 m) dome in your backyard to get a feel for it.

CHAPTER 12

Illustrated Guide to Dome Construction

How We Built the Honey House

At the time of writing, our personal experience with dome building consists of the construction of the Honey House. The Honey House is a 12-foot (3.6 m) interior diameter earthbag dome sunk two feet (0.6 m) into the ground. We did the bag work in 19 days (with a lot of head-scratching) over a period of two months. We averaged a crew of three to five people working modest five-hour days in between other work schedules. A professional backhoe operator did the excavation work, and all the reject dirt used in the bags was delivered. Exterior plastering and sculpting went up in a few big parties; the fine detail work lasted throughout the summer. All in all, we have spent about $1,500 to date, which includes the windows and custom-made wood door. The Honey House was built below permit size in our backyard in the middle of town.

We learned a lot about the overall process, dynamics, structural integrity, and limitations of building a dome with earthbags. What we offer is a narrated sequence of events that depict the process of the construction of the Honey House. We include a few parallel design options along the way to show how the dome can be adapted to different styles and climates. The structural principles will remain the same, with our focus on building a modestly sized,

self-supporting earthbag dome. Think of this as a montage from a movie rather than a step-by-step instructional video.

PRIOR PREPARATION CHECKLIST

- Do tests to assure quality of earth for dome building and determine proper moisture content.
- Practice building a small wall project first to become acquainted with the basics of earthbag building before taking on the more intricate work of building a dome.
- Make scale drawings and/or build scale models to get an idea of what you want to build.
- Build *all* forms needed for the project. Make sure they are sturdy, and deep enough to accommodate the corbelling process.
- Prepare all the strip anchors and/or Velcro plates for the installation of electrical boxes, shelving, steps, eaves, etc. Having these pre-made keeps the building momentum going.
- Make and assemble all needed dirtbag tools: tampers, sliders, cans, compass, etc.

Corbel Simulation Test with Tubes

To get a feel for the corbelling process, let's corbel a few rows of ten-foot long (3 m) tubes on the ground. Practice getting a "feel" for how tightly you can bend the tube into a curve. Tamp the first tube and measure the finished width and thickness (refer to Chapter 3 for techniques on laying tubes). You can omit the barbed wire for this test. Measure three inches (7.5 cm) in from the outside edge of the tube and draw a line on the tube for its full length. This line indicates how far the next row will be stepped in (corbelled). Fill and lay the next tube up to this marked three-inch (7.5 cm) line. Tamp down this second row and measure the thickness of this row. Did it get much flatter than the one below? Now try a third row, stepping it in four inches (10 cm). Tamp it and measure its thickness. If the rows tamp down significantly flatter the more they are stepped in, it's likely you have the right ingredients for following the compass recipe we have provided. Four inches (10 cm) is about the maximum that we feel comfortable over-hanging (corbelling) a tube or bag (Fig. 12.1).

The stronger the corbel (that is, the further the rows are stepped in), the more likely the earthbag will be flatter than the preceding rows. The reason is that tamping forces more of the material into the part of the bag that has the least resistance, the part overhanging the previous row. Keep in mind there is a definite limit to the amount a row can be stepped in. This is determined by the width of the bags/tubes, the characteristics of the fill material, and the quality of the work being performed. Because of the soil, materials,

skills, and other unforeseeable conditions that may present themselves, you as the builder/architect will have to adjust your design based on real life circumstances, rather than on what you have read in this book.

We used a variety of bag sizes to construct the Honey House, using the larger bags (way-too-big and 100-lb. bags) down low where they were easier to work with, and finished the dome off with narrower 50-lb. bags and equivalent-sized tubes. When building earthbag structures, and especially domes, use your largest bags near the base of the structure and progress to smaller bags as the walls increase in height. We want to distribute the weight of the whole building so that wider bags support the base that carries all the compressive force, while progressing to narrower and lighter bags towards the top.

Drawings for the Honey House

(Refer to Chapter 11 for detailed explanations on making architectural drawings for domes.)

The first drawing we made was an "elevation" sketch. This is essentially a cross-sectional drawing of the height and width and shape of the dome with foundation details and floor level indicated (Fig. 12.2).

This shape closely resembles a catenary-shaped arch (see Chapter 10 for more about arch shapes).

The second drawing is called the "floor plan." This is a horizontal cross-section showing the thickness of the walls with any buttressing included.

This plan also shows window and door placement and sufficient wall space between these openings (Fig. 12.3).

12.1: Corbelling simulation test done on the ground with 15-inch working-width tubes. Get a feel for the corbelling technique and test the behavior of your soil.

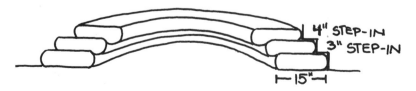

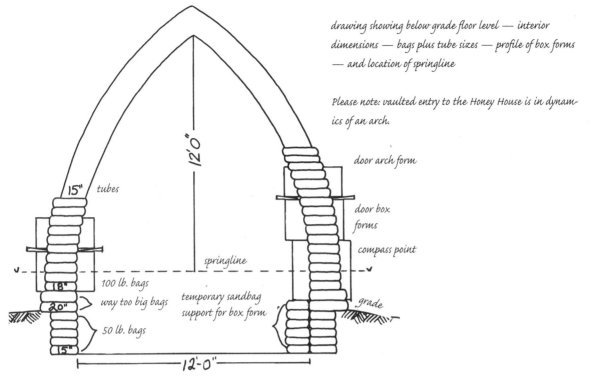

drawing showing below grade floor level — interior dimensions — bags plus tube sizes — profile of box forms — and location of springline

Please note: vaulted entry to the Honey House is in dynamics of an arch.

door arch form

door box forms

compass point

grade

12'0"

15" tubes

springline

100 lb. bags

way too big bags

temporary sandbag support for box form

18"

20"

50 lb. bags

15"

12'-0"

12.2: *Elevation drawing depicts below-ground foundation, internal dimensions, and compass profile.*

Include a third drawing of any electrical, plumbing, shelving attachments, or extended eave details that will be installed in the proposed dome.

Once these drawings are completed, you will have a better idea of what the dome will look like and what it will require. Sometimes, though, it may still be difficult to picture what the finished project will look like. If this is the case, consider building a clay model of the proposed structure.

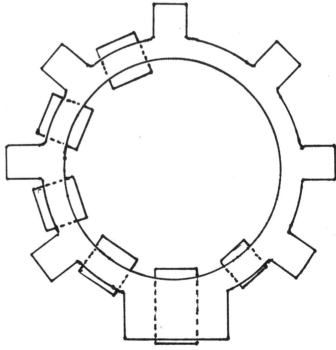

12.3: *Top view of building shows buttressing placement and door and window box forms.*

Building Clay Models

Since domes are three-dimensional, it is easier to comprehend their design in a three-dimensional medium, like sculpture. We use a screened and sandy, clay-rich soil mixed with sun-dried short grass clippings, but any short fiber will suffice for sculpting models. We add a little borax to the mix to inhibit mold growth.

Blocks of wood cut into an arch shape make dandy arched window and door forms for your model.

A cardboard cutout in the shape of the dome gives us our interior compass radius. The bottom center of the cardboard is pegged to a plywood platform that enables it to rotate (Fig. 12.4a & b).

The mud follows the shape of the cardboard compass over the window and door forms until the roof is completed. Voila — home sweet dome! (Fig. 12.4c & d).

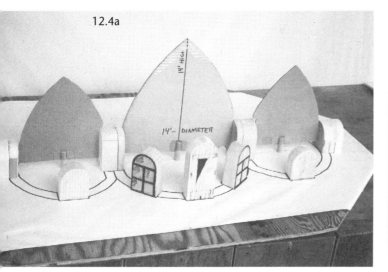

12.4a

12.4b

12.4c

12.4d

Dome Building: Sequence of Events

We are assuming that you have carefully read the step-by-step earthbag wall building chapter to acquaint yourself with the Flexible-Form Rammed Earth technique we've been going on so much about. If you have followed the Prior Preparation Checklist, earlier in this chapter, you should have all your drawings completed, model making done, soil and bag tests finished, forms, strip anchors, Velcro plates, and all the needed dirtbag tools built and ready. If this is the case, you are ready to begin.

Excavation

Let's begin the excavation by locating the center of the dome. Drive a stake or post into the ground at that point. Attach a non-stretchy rope or light chain to the center post so that it can rotate easily. Make sure it is long enough to reach beyond the proposed radius. Mark the rope or chain at least two feet (0.6 m) beyond the interior radius to include the width of the wall (Fig. 12.5).

That is to say, if you are building a 12-foot diameter (3.6 m) interior dome, make a mark on the rope or chain at eight-foot radius (2.4 m). This gives

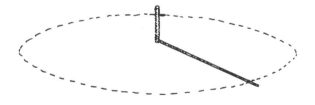

12.5: Upper drawing caption: A rope or chain attached to a stake in the ground delineates the center of the diameter. Lower drawing caption: Allow extra width for installation of waterproof membrane or insulation.

you the approximate location of the exterior walls. Make this circle big enough to work around comfortably, especially if you are installing rigid insulation below grade. Extra space is easily backfilled later, once the bag work is brought up to grade level. Using powdered lime, chalk, or some other non-toxic marking material, draw a complete circle on the ground as you rotate the tautly stretched chain in a full circle.

When beginning the excavation, remove the topsoil and set it aside to be used later for a living roof or landscaping. This soil is full of microorganisms, weed seeds, and a host of organic matter that is unsuitable for putting inside the bags. Humus belongs in the garden, not in the walls of your home. This topsoil can be anywhere from nonexistent to eight inches (20 cm) or more thick, depending on your location.

Remove the remaining soil from the building site to the desired depth. Perform tests on this soil (described in Chapter 2) to determine its suitability for use in the bags or, later, as a component of an earthen plaster, depending on its clay content. Separate and pile usable building and plaster soils in convenient locations. Pre-wet the building soil to optimal moisture content and cover with a tarp to protect from rain, sun, and windblown debris.

Compass Installation

At this point it's time to install the building compass that will be used to delineate the shape of the dome and to provide a guide to help maintain level as the wall rises. Pull up the stake you used to mark the shape of your circle and replace it with the compass pole. (A complete treatment of how to set the pole, what the different parts consist of, and how it works, is contained in Chapter 3 under the heading "Construction-Size Architectural Compasses for Domes and Round Vertical Walls").

When the center pole is set plumb, and the horizontal arm with the angle bracket is attached, check the excavation with the compass to make sure the hole or trench is correct. You may have to trim the excavation a little to be certain the circle of the excavation matches the circle scribed by the compass arm. It's much easier to trim the hole to the right circumference than to reset the compass.

Foundation and Stem Wall

We began our bag work right on the ground of this excavation; these bags became the foundation of our structure. The first row of bags can be filled with gravel to inhibit capillary action from the ground up into the earthen walls. The continuing bag work up to grade consisted of the earthen fill we had prepared previously (Fig. 12.6).

12.6

Use the compass arm to delineate the shape of the structure. The angle bracket set on the horizontal arm denotes where the inside circumference of the finished, tamped bags should be. Use this arm when initially placing each bag. Set the filled, untamped bag about one inch (2.5 cm) outside this angle bracket. This will allow for expansion of the bag once it is tamped. The horizontal compass arm must be used for each bag placed around the circle (Fig. 12.7).

12.7: *Use the compass arm to delineate the contour of each bag to conform to the circle.*

When this first row of earth-filled bags have been placed and tamped, swing the compass arm around the perimeter to see how well you have done. You may need to further tamp a few bags to meet the angle bracket on the horizontal arm. Or possibly the bags came in too far and need to be pounded outward a little so the angle arm can pass without binding on the bags. Adjust the bags, not the arm. This first row will tell you how the following rows need to be modified to adjust to the proper diameter of the compass arm (Fig. 12.8).

If you read the section in Chapter 3 concerning the building compass, you will have bound or taped a level onto the horizontal arm of the compass. As you rotate the compass arm around to check the

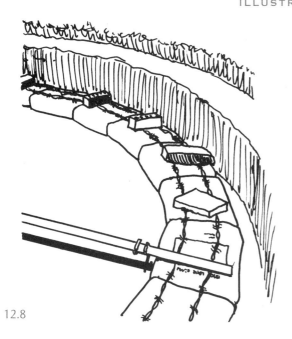

12.8

12.9: *Continue the bag work until the bags are up to or just below grade. This is where the stem wall of the structure will begin.*

proper placement of each bag in the circle, also check the level of each bag in respect to the other bags. The idea is to keep each row as level as possible. Measure the height of the bags after they have been tamped. The bags we used at this stage of the building process tamped down to a thickness of five inches (12.5 cm). Once the average thickness is determined, it's simply a matter of raising the horizontal arm of the compass the corresponding amount for the next row. Tamp the bags until they are level with this setting.

Several options may be considered for creating the stem wall. Two rows of concrete, stabilized earthbags, or gravel-filled tires, are all effective stem wall options. If using exterior rigid foam insulation, install

the foam high enough to protect the stem wall. Another option at this point, instead of insulation or in addition to it, is to install a moisture barrier around the perimeter of the bag work up to the top of the stem wall. (All of this is covered in excruciating detail in Chapter 4).

For the below-ground bag work that we are describing, the dome walls are designed to be vertically plumb (like a yurt, or what we refer to as *kiva-style*) to provide a little additional interior height. Think of it as a *knee wall* in an attic. Choose a stem wall height that is appropriate for your climate - usually the wetter and colder the environment, the higher the stem wall. Your individual circumstances are paramount in making these choices (Fig. 12.9).

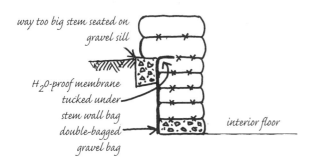

way too big stem seated on gravel sill

H_2O-proof membrane tucked under stem wall bag

double-bagged gravel bag

interior floor

12.10: *Honey House foundation detail.*

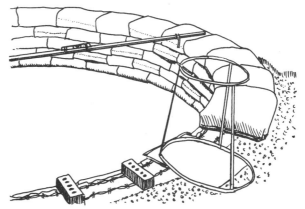

12.11: *Way too big bags on gravel sill at grade.*

For the Honey House dome, we began the below-grade bag work using a standard 50-lb. bag that tamped to about 15 inches (37.5 cm) wide. For the stem wall we switched to the larger bags we call way-too-big. We maintained the same interior diameter of the bags below, allowing the way-too-big bags to extend beyond the outside perimeter of the lower bags. We accomplished this by backfilling and tamping the outside space below grade (after installing a moisture barrier) with gravel, up to the level where the stem wall bags begin (Fig. 12.10 & 12.11).

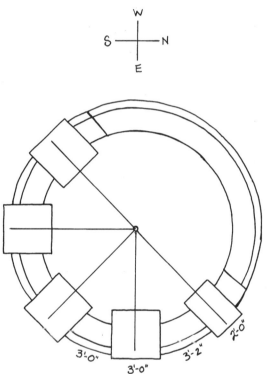

12.12: *We made a top, or plan view, drawing to delineate how we wanted the windows and door oriented by sighting the compass arm down the center of the box forms.*

Installing Door and Window Forms

After completing one full circle of way-too-big stem wall bags, we installed the door form, a three- by three-foot (90 by 90 cm) rigid box form. We used the compass to designate exactly where we wanted to position the door. We set the lower portion of the box form on the wall, raised the horizontal arm on the compass to the same height as the box form, and squared it down the center of the box (Fig. 12.12). With a marking pen, we drew a line on either side of the box right on top of the tamped bag wall. The next row of bags was begun on either side of this form, to anchor it evenly, and then the next row of bags was completed.

We used way-too-big bags that tamped out to about a 20 inch (50 cm) working width as our first two rows of stem wall bags, giving us about a 12-inch (30 cm) high stem. On top of this second row of way-too-big bags we set up our window forms, each one being three feet wide and two feet high (90 by 60 cm) (Fig. 12.13).

Hold off on installing the barbed wire until these forms are set in place. It's easier to set up the box forms directly onto the tamped bags.

When measuring the solid area in between window and door openings, use the exterior wall measurements for the calculations. In a circle, the

12.13: *Use the compass to align box forms.*

outside of the wall will, of course, be wider than the inside measurement. In order to determine the best way to fill the space between the box forms with bags, make the best use you can of the length of bags being used.

Buttressing

To ensure there was enough bulk in between the window openings, we added buttressing, as well.

As this is a small structure with compact openings, the relatively narrow space between the windows did not compromise the integrity of the structure. For a larger dome, or if the windows are to be installed at a higher level (which would be the case if we were building a dome with a floor at grade level), the window boxes should be further apart to increase the width of the solid wall between them (Fig. 12.14).

Now we can lay barbed wire in between the forms, being sure to incorporate any buttresses.

We used 100-lb. bags around our door and windows for extra mass. Once we got all our forms locked in place, we continued the bag work around the entire circle and tamped the whole row, making sure all of our box forms were secure, and lay barbed wire over the surface of the tamped wall and in between the forms (Fig. 12.15 & 12.16).

great example of where to use a scab!

12.14: *Incorporating buttresses is another way to increase the mass in between openings while keeping the openings relatively close together.*

12.15 (below): *Extend extra-long strip anchors beyond profile of wall as a built-in ledge to support a dormer made from sculpted cob, bamboo, fired brick, etc. Note: Buttressing extends into interior of dome.*

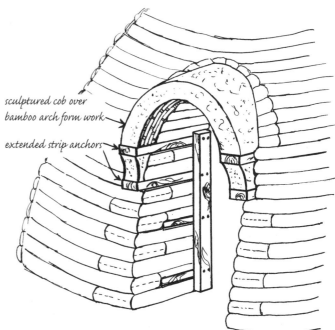

sculpted cob over bamboo arch form work

extended strip anchors

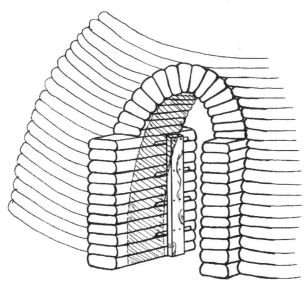

12.16 (left): *Door buttressing can be built out of tubes or bags that extend into the interior, exterior, or both. The door opening is the weakest part of the wall in a dome. To ensure structural integrity, incorporate buttressing alongside the door forms. Buttressing provides extra wall width needed for installing a plumb doorjamb. (The shaded area indicates the contour of the corbelled wall.)*

UPTIGHT TIP FOR BETTER COMPRESSION FIT

Install the bags up against the window and door forms as tightly as possible. The forms are an integral part of the structural dynamics of the dome during the construction process. They act as a link in the chain of each ring of bags or tubes. They are not there just to make a hole; they are there to *create* a whole — a whole double curvature, monolithic structure. Each row needs to think that it is making a complete ring, an unbroken circle, a full connection, mini-tension ring, bond beam, whatever you want to call it. Without the connection of a unifying compression fit on either side of every form, the dome could collapse.

Remember to interlock any tubes with the bags up against the forms, rather than ending the tubes against the forms. We prefer to use bags around window and door forms as they create a much tighter fit than a tube and offer greater compression.

The Springline: Corbelling the Bags and Tubes

Using the drawing as a guide, adjust the compass inward to match the calculation for the first row to be corbelled. In our case, we shortened the compass arm one-half inch (1.25 cm). This corresponds to a total step-in of one inch (2.5 cm) for the full diameter of the dome. As is shown in the drawing, the profile of the dome steps in very gradually at first — almost imperceptibly (Fig. 12.17).

That's OK; it's a cumulative kind of process. At this point we began laying coils (tubes) for the walls, in conjunction with individual bags up against the forms.

With the compass adjusted to accommodate the first step-in, we loaded a 20-foot (6 m) tube chute, tied off at the bottom from the inside (see Chapter 3). Before laying this coil, it's best to go ahead and do the bag work around the door and window forms first and any buttressing. Get all the bags gently secured with mild tamping. Then begin laying the coil as snugly as possible up against the bags. After laying the coil, tamp it, and all the bags around the forms, until hard (Fig. 12.18).

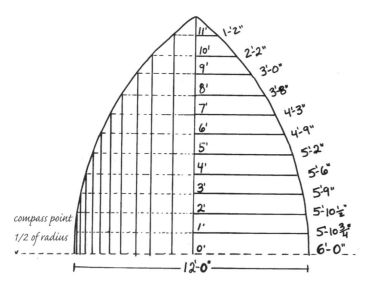

12.17: *This diagram shows the approximate lengths to to which the compass arm radius must be adjusted at every one foot of height. You can draw a larger diagram that includes measurements at every one inch (2 cm) or less for at-a-glance reference calculations during construction.*

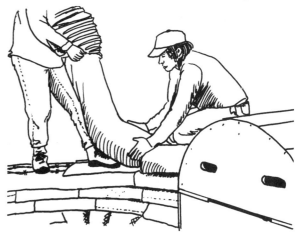

12.18: *Corbelling the roof is done by stepping in each row the specific amount determined by the compass profile.*

The trick to laying a coil is to have plenty of dirt available to fill the tube quickly so as not to tire out the person holding the tube chute — your human bag stand. We also found it helpful to place sliders under the feet of the human bag stand, to keep them from sticking to the barbed wire, and to make it easier to position the tube accurately according to the compass.

Our goal was to step-in the rows, but to leave enough of a gap so that when they were tamped we would still be able to swing the compass arm around without it getting stuck on a slightly protruding bag. It's a good idea to avoid having to smack the bags outward to make the compass fit, as this is more apt to disturb the compaction of the enclosed dirt and loosen the fabric, deform the shape of the coil, and decrease its woven tension under compression (Fig. 12.19).

From this point on the procedure remains the same. Adjust the compass for the next row. Install any needed strip anchors next to the forms. Install barbed wire. Install bags around box forms, remembering to hard-ass these bags if not installing strip anchors. Lay coils in manageable lengths that will permit you to overlap the seams of the previous row. Fit the ends of coils tightly against each other by adding a couple of extra cans of dirt and shaking the dirt down well. Twist the end of the fabric tight and neatly tuck it under itself. Tamp the whole row. Repeat this process until you get up to the height where the arch forms need to be installed.

Installing Arch Forms

Placing the arch form on top of the box form is a relatively simple process (as explained in Chapter 6). Install the arch forms deep enough towards the inside of the dome to accommodate the angle of the corbelling process. Use the compass arm to level the arch form along its long axis, the same way it was used to level the box form below it. Use a level along the width of the base of the arch form to level it from side to side. The wedges used between the box and arch forms can be tapped in or out in all directions to achieve this level. Once you are happy with the level of this form, begin the next row, which will also involve the *fan bags* (Fig. 12.20).

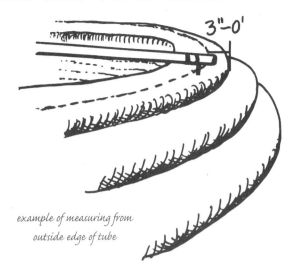

example of measuring from outside edge of tube

12.19: *Later, as the rows were stepping in three or more inches, we used the compass to draw a line on the outside edge of the previous row, and used the line and the compass as our guide for stepping in the following row.*

12.20: *Incorporate the barbed wire from the wall in between every fan bag around the arch forms, as shown in the step-by-step illustrated guide, and integrate tubes with bags against the forms.*

The fan bags around the arch forms for a dome are built pretty much the same as they are in our step-by-step wall-building chapter. The main difference is that, because the slope of the dome roof is stepping in every row, so too will the fan bags (Fig. 12.21).

Continue laying coils up against the fan bags. The step-in increases as the profile of the dome curves increasingly inward.

12.21: *Placement of the inside edge of the fan bags should be aligned with the compass.*

If there is any wobble at all, check to make sure the bags and coils are snug up against each other and the forms. If the dirt is either too wet and soggy, or too dry, or if the moisture throughout the mix is inconsistent, the soil will not compact properly and will feel unstable. For safety's sake, take the time to customize the soil accordingly (Fig. 12.22).

As the dome gets taller, straw bales placed around the perimeter with boards on top make a simple scaffold that contours to the shape of the dome and protects the bags from UV exposure.

Be sure to check periodically to see that the compass pole is still rigid, as the higher the horizontal arm is raised, the more torque is placed on the vertical pole, pulling it out of alignment. Stacked sand bags around the base help steady it. So does adding rope or wire tie-offs at a level over head height (Fig. 12.23).

If the center pole is not long enough to reach the entire height of the dome, it can be extended by adding a *coupling* (a double-ended sleeve) that fits over the top of the center pole, and another pole can be set into it. Duct tape wrapped around the pipes helps create a snug fit.

12.22: *After tamping, each row should feel as secure as a sidewalk.*

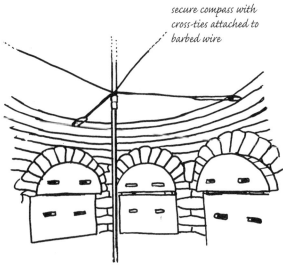

secure compass with cross-ties attached to barbed wire

12.23: *You will definitely need to shore up the compass pole with cross-ties if a coupling is used. Raise these cross-ties as the wall gets higher, to add more stability.*

Second-Story Floor Joists

If you are planning to install a loft or second floor, wait until there are at least one or two rows of tubes over the tops of the finished arch windows. Two rows are better, as the seams can be staggered to help create as much of a tension ring as possible, although the door may still be under construction.

A second story makes a great interior scaffold as well, even if built solely for the construction process and sawn off later. Except for a few more rows to finish the door, all bag work around the first floor windows has now been completed (Fig. 12.24).

Optional Eaves

Roof eaves can be installed at any desired height above the finished windows, but on top of at least two complete locking rows of tubes. Extended eaves also make handy built-in exterior scaffolding, and can be installed as an extension of the second story floor space or independently at a different height (Fig. 12.25a & b).

If you've decided to install extended eaves, you want to place them evenly around the perimeter of the wall. A simple formula that will help calculate the

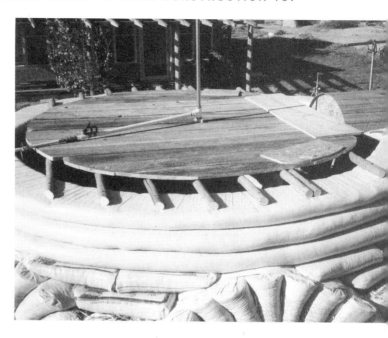

12.24: *We used lodge poles for the joists of our loft. We ran them straight across and staked them into the wall with 12-inch (30 cm) spikes. A more sophisticated method could be any of the systems described for anchoring rafters or shelf brackets on top of Velcro plates, as described in Chapters 7 and 9.*

exact distance to set the eaves from each other can be found by figuring out the circumference of the outside of the wall (this and other formulas for making calculations involving circles are found in Appendix C). Use the horizontal compass arm to determine the radius from the outside of the bags to the center of the pole

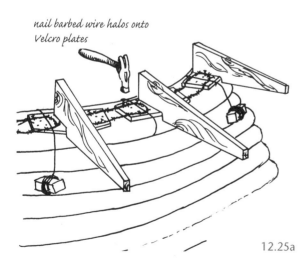

nail barbed wire halos onto Velcro plates

12.25a

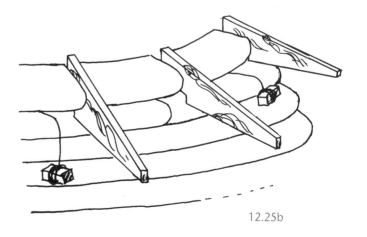

12.25b

12.25a & b: *(a) Extended eaves attached to plate and Velcroed into tamped bags. (b) Infill between eaves with the next row of bags and tamp snugly.*

at the desired height of eave installation. To make the job of installing eaves easy and accurate, use the compass arm. You can also use the compass for aligning second story loft joists that radiate from the center like a spoked wheel (Fig. 12.26).

Dome Work Progresses

Once the second story is up and a couple more courses of tubes are laid, the progress of the dome construction accelerates quite rapidly as the step-in becomes more pronounced. As the wall height increases, each new row of tubes becomes shorter. This type of construction is easiest to do with a synchronized crew (Fig. 12.27 & 12.28).

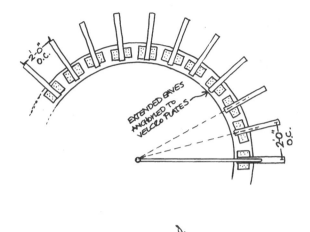

12.26: *Use the compass to align position of extended eaves or second story floor joists. (Top view.)*

12.27: *Laying barbed wire at this height is easiest done with the reel safely located on the ground.*

Of course, this would be a great time to have a friend with a backhoe or bucket loader, but it isn't compulsory, just convenient. We actually finished off the top of the Honey House with only three people because we had become very proficient can tossers by then. Believe us, by this point in the construction, your proficiency will have increased noticeably too.

When we were about two feet (0.6 m) above the second story loft we started thinking, "Wouldn't it be nice to see the mountains from up here?" We used the compass to align two 3-foot (0.9 m) deep by 2-foot (0.6 m) wide arch forms. Window forms at this height will need to be plenty deep as the roof steps in so dramatically. We continued to use the wider 100-lb. fan bags around the arch forms. Set the form as much as possible towards the inside of the dome and support it underneath with stacked cinder blocks or scrap wood bracing (Fig. 12.29).

12.28: *Cooperative teamwork involves two people on the wall, one to be the human bag stand and the other as the loader. A person on the ground supplies the dirt and another, on the ladder, acts as the intermediate can-tosser.*

Corbelling Safety Tips

A good rule of thumb when stepping in bags/tubes, is to step-in only a maximum of one-quarter the width of the working bag. For example, if you are using tubes or bags that tamp out to a 12-inch (30 cm) width, they should only be stepped in a maximum of three inches (7.5 cm). These are all approximations, and adjustments can be made depending on the soil used and your own comfort with the process. The bags/tubes we used for the Honey House tamped down to 15 inches (37.5 cm), and we were able to step them in a maximum of four inches (10 cm), a little more than the one-quarter rule, but our soil mix was primo (Fig. 12.30).

If you are having trouble following the curve of the profile, back off to a more gradual step-in pattern. It is safer to accommodate the limitations of the materials, even if it means gaining a foot (30 cm) or so of extra height, than it is to risk compromising the structural integrity of the building by trying to force the material to comply with a drawing. In this case, the profile may need to be a more gradual curve

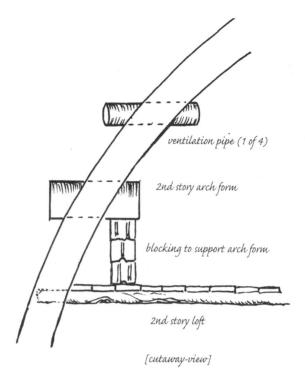

ventilation pipe (1 of 4)

2nd story arch form

blocking to support arch form

2nd story loft

[cutaway-view]

12.29: *Providing support for the second-story arch forms may require extra bracing. Note the installation of upper ventilation pipes.*

12.30: *Note how far the upper row of bags overhangs the preceeding row.*

drawn free-hand, or with a drawing compass (Fig. 12.31).

Closing in the Dome

Naturally, the circle becomes smaller as it gets closer to the top. As a result, it becomes harder to lay the tubes in a tight circle. When we were about five rows from the top, we switched back to 50-lb. bags. Using bags freed up our hands, making it easier for us to close in the dome with only three people. We were also able to custom contour the shape of each bag to fit a tighter circle. The pattern evolved from an eight-

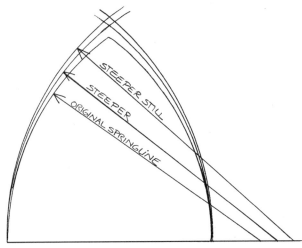

12.31(above): *Steepen the profile by drawing several lines from a compass point further from the center of the diameter, and make new calculations from the steeper profile.*

12.32(left): *A halo of barbed wire Velcros the bags together, while the chicken wire provides a grippy surface for the later application of plaster from inside.*

pointed circle, to a six-pointed circle, to a square with four bags, to a triangle-shaped hole formed by three bags. We threw some chicken wire over the opening, added a halo of barbed wire, and laid two bags on top. A single bag (with barbed wire under it) was laid over the seam created by the two lower bags and we called it done. It was actually a little anticlimactic, but satisfying and fun just the same (Fig. 12.32).

That's it! The process is relatively simple and repetitive, which frees your mind for creative problem-solving when and if problems arise. It is impossible for us to address all the things you may encounter, but with a good basic understanding of all the elements involved, a willingness to experiment, and the ability to adapt your expectations to the reality of the medium, you too can build a uniquely beautiful structure that will outlast any contemporary building currently endorsed by conventional construction (Fig. 12.33).

12.33: *The completed bag work of the Honey House after removal of the forms.*

Directions for a Dome With a Grade Level Floor

If you use a rubble trench or some other foundation system in order to have a floor at *ground level*, you will begin your bag work on this foundation (Fig. 12.34).

Build the stem wall and place any door forms for this structure as described previously. You will probably want to install your window box forms 28-30 inches (70-75 cm) above the floor level, or at about normal counter height.

Use 100-lb. bags, or way-too-big bags, or two rows of 50-lb. bags side-by-side as a vertical wall up to the height at which the window box forms will be installed. Set up your box forms and anchor them in with one more row of bags. At this point begin your springline, and switch to single width 50-lb. bags or equivalent width tubes. Continue to integrate the 100-lb. or way-too-big bags around the door and window forms, as described earlier, for extra bulk around the openings (Fig. 12.35).

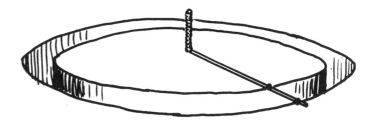

12.34: Use a string compass to designate interior and exterior of foundation trench. Trench style foundation for a grade-built dome, include trench work for any buttressing. Fill entire trench with rubble rock and gravel up to or just below grade.

12.35: Grade-level dome.

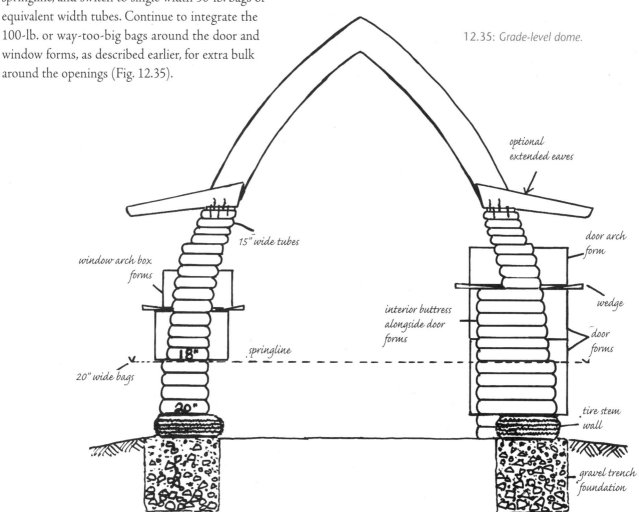

optional extended eaves

15" wide tubes

window arch box forms

door arch form

wedge

interior buttress alongside door forms

door forms

springline

18"

20" wide bags

20"

tire stem wall

gravel trench foundation

You can also include buttressing, if you intend a sculpted gutter system. In other words, whatever it takes to provide ample mass at the perimeter of the walls, so that you can safely begin the springline two feet (0.6 m) or so higher than grade level. Continue all bag/tube work as described in this chapter to complete the dome.

Are Buildings for Squares?

We have systematically turned our natural resources into modular components for building uniform compartments. No wonder the book that dictates our construction practices is called the Uniform Building Code.

To follow the Uniform Building Code requires uniform building materials. We turn round trees into dimensional lumber. We cast cement into specific size blocks. We have learned to build with products instead of processes. I-beams, plywood, brick, and lumber are all products designed for corners instead of curves. We do this not because it makes stronger buildings; we do it to support the manufacturing industry. Square building materials are easier to stack and transport than curved ones. Rectilinear, square box-itecture promotes the consumption of square building products.

When we switch to a round construction mentality, we discover we are in opposition to the status quo. It is difficult to integrate square products into round structures without modifying them significantly or creating waste. So we look to alternative materials to build with: adobe, cob, stone, straw, paper, cordwood, timber, bamboo, tires, rammed earth, wattle and daub, feedbags, barbed wire, and other materials outside the rectilinear shelves of our local hardware store.

Alternative builders are shopping at agricultural supply houses, scavenging at dumps, salvaging recyclable material and turning waste products into creative processes, making mansions out of mud holes, and developing sustainable systems that build sustainable societies.

Besides, the curve is coming back into fashion. Our technology reflects this trend toward curves. Cars are becoming more aerodynamic. Computer consoles, kitchen appliances, sports equipment, and cell phones are getting more sensual. A boom box is no longer a box and the Super Bowl is not the Super Box.

CHAPTER 13

Roofing Options for Domes

The exceptional strength of a corbelled rammed earth dome can easily be designed to carry the weight of a nine-inch (22.5 cm) thick *living thatch* roof (we've done it!) or a hefty layer of sculpted adobe. Traditional thatch, terra cotta tiles, lime plaster over adobe, mortared slab stone, slate, wood, and even asphalt shingles (a good, cured rammed earth will hold long roofing nails), are also suitable roofing materials.

For insulating a corbelled dome in a cold climate, we have designed a wood frame system attached to eaves that are installed during construction and then insulated, sheathed, and shingled. The finished roof resembles a roof-sized bell. In a warm, frost-free climate, lime/cement stucco sculpted into a spiraling gutter system lined with mosaic tiles can be a stunning way to collect precious rainwater.

13.1: *The Honey House receiving a natural-clay rich earthen roof plaster.*

Earthen Plaster Roof for Dry Climates

Layers of a quality clay plaster built up to at least 6 inches (15 cm) thick make an excellent protective covering for an earthbag dome. Good quality clay has natural water resistant characteristics. It should be sticky and fairly stable. We mix this nice, sticky, stable clay with about a 60-70 percent sandy soil, like our reject sand (see Chapter 2 for more about reject sand) with a copious amount of long straw right out of the bale, 6 to 12 inches (15-30 cm) long (Fig. 13.2).

This base layer goes on a lot like cob. The main difference is the higher clay ratio. When clay gets wet it swells, inhibiting further water migration. The first one-half to one inch (1.25-2.5 cm) will soften and the finest clay particles near the surface will wash away exposing a gross network of deeply embedded straw.

Over this clayey four- to six-inch (10-15 cm) base coat, a second plaster coat gets smacked on about three quarters to one inch thick (1.875 cm-2.5 cm). A strictly natural earthen roof plaster will require periodic replastering, but the steeper the slope of the dome, the quicker it sheds water. The quicker it sheds water, the quicker it dries out. Steeper domes translate into less plaster maintenance.

Something to think about when designing an earth-plastered dome is how to protect the windows from all the silty water that comes off the roof. We built dormers and sculpted rain gutters out over the arched windows and down the buttresses — this design also directs water away from the foundation. This functional aspect became its most pleasing esthetic attribute. Isn't it just like nature to combine function with form? (Fig. 13.3).

Lime Plaster Roof

In a low rainfall climate like Tucson, Arizona, a lime plaster over an earthen base coat is a traditional method of protecting adobe dome roofs. Lime plaster can be sprayed on with a mortar sprayer, applied by hand, or troweled on over a thick, rough earthen base coat. Multiple layers of lime plaster followed by several coats of lime wash offers more protection than simply one or two thick coats of lime plaster. In a dry climate, periodic lime washes are enough to protect a lime-plastered dome.

13.2: *For a water-resistant earthen-base plaster, we increase the percentage of clay to sandy soil, pack it full of long straw, knead it into hefty loaves, and smack them onto the dome.*

13.3: *The vertical faces of our Honey House are protected with lime plaster, and the exterior windowsills are made of lime-stabilized earth. Both applications have held up very well for five years.*

Waterproofing Additives for Lime Plaster

We personally have not experimented with water-proofing additives for lime plaster, but there are a few that have been shown to work well. Adding Nopal or prickly pear cactus juice is said to aid water resistance without compromising lime's ability to transpire moisture. The cactus pads are cut up and left to soak in a barrel filled with water until the mixture ferments. The resulting slime is strained and then used as 40 percent of the total water to make lime plaster or lime washes.

A mixture of alum and water and a mixture of soap and water are alternately mopped onto the surface of a lime-plastered structure. This method of water resistance has been used effectively in Mexico for decades. Usually applied to parapet tops, horizontal surfaces, and domes, several alternating coats will give ample protection for many years. Lime with pozzolan is yet another water resistant strategy to explore.

Tile and Flagstone

A coarse lime or cement plaster makes an excellent bonding surface for tile, particularly thick tiles that can be set in nice and deep. Thick broken Mexican tiles and flagstone are complementary. A roof of flagstone with rain gutters lined with colorful tiles would be a stunning way to protect a dome in the tropics or the desert (Fig. 13.4).

Stabilized Earth

This dome is constructed of six-inch (15 cm) thick rigid foam protected with a four-inch (10 cm) thick layer of sculpted cement-stabilized earth by artist/sculptor Robert Chappelle. What is truly remarkable about this home is that it is in central Vermont! (Fig. 13.5).

After much experimentation, Robert mixed the optimum ratio of cement (a hefty 16 percent) into his sandy reject soil, resulting in a plaster that has withstood the ravages of Vermont winters since 1994. His sculptures continue to endure, he says, unchanged since the day he completed them in the early 1990s.

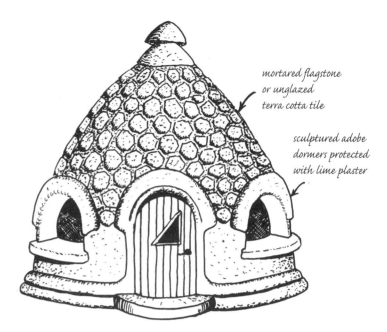

mortared flagstone or unglazed terra cotta tile

sculptured adobe dormers protected with lime plaster

13.4: *Low-fired clay tiles set in earthen mortar have been a traditional roof covering on domes in the Middle East for centuries. The mud mortar allows for transpiration, while the tiles inhibit erosion.*

13.5: *Robert Chappelle's home in Vermont. This is not an earthbag dome!*

Keep in mind, however, that his structures are thick rigid foam, not raw earth, and therefore less likely than living earth to be affected by changes in weather (Fig. 13.6).

13.6: What fascinates us is the effectiveness of his stabilized earth mix in such a harsh climate. Photo credit: Robert Chappelle

13.7: Keep it alive, not stabilized!

Robert started out doing a four-inch (10 cm) thick single coat application, but recommends doing several thin one-inch (2.5 cm) thick coats instead. Thin multiple coats have proven to remain crack free and water resistant. His success inspires us to experiment with trying lime stabilized soil (instead of cement) as a protective covering over earthbag domes, as lime is more compatible with the vagaries of raw earth.

Ferro Cement

For someone, somewhere, cement may be the most appropriate resource for protecting a dome. In the Bahamas, where temperatures remain above freezing and cement is more available than clay, it would be a relevant choice for plastering.

The primary factor to consider when choosing cement-based plaster over earthbags is to fill the bags with a coarse, sandy soil low in clay. This mixture is less apt to be affected by internal moisture or expansion and contraction. Adding a percentage of lime to the cement-based stucco is helpful for increasing the plaster's ability to transpire moisture while limiting the migration of external moisture from entering. There are several excellent sources to learn more about Ferro Cement. These are available online and in books.

Living Thatch Roof for Domes

We were once told that it would be impossible to put a living roof on a pitch as steep as an earthbag dome, but with all the little corbelled steps, the layers of built up earth stayed put just fine. We applied the living roof in two layers. The first layer was a cob mix of mud, sand, and straw about six inches (15 cm) thick. The second layer was a "living cob." To this cob mix we added live Bermuda grass roots. A mist sprinkler was installed on the roof, but thanks to El Nino we were blessed with three weeks of rain. The material never slid; instead it sprouted in three days (Fig. 13.7).

We did not use any sort of waterproof membrane. We simply stacked the cob up on the steps created by the corbelled tubes. This resulted in a 23-inch (57.5 cm) thick roof composed of 14 inches (35 cm) of tamped earthbags and about nine inches (22.5 cm) of cob mixes. We sculpted cob gutters above the arch windows and down the buttresses, which directed water away from the foundation. We had two small leaks from too much concentrated moisture when we started using soaker hoses for irrigation — we have an average rainfall of only about eight inches (20 cm) per year; a living roof is more suited to a moister climate, but we just had to try it! In a wetter climate we'd use a layer of bentonite clay over the bags as a waterproofing membrane, or stagger sheets of a thick EPDM pond liner. Then throw a polypropylene fishing net over that for the live cob roots to anchor on to (Bentonite clay can be used successfully as a waterproofing agent for underground houses according to experiments done by the University of Minnesota's architectural department). We'd also add extended eaves, and let nature do the watering. We chose a hybrid Bermuda grass for its incorrigible root system and long, dense, droopy appearance. We imagined it growing into a living thatch that would stabilize the earth while helping to shed water. No mowing required.

Our experiment led us to realize that one can successfully grow grass on a steeply pitched roof. The roof aids in holding the sod, as it gets wider from top to bottom. It creates the same effect as trying to pull a knitted cap down over your head. Also, the structural dynamics of a properly designed corbelled rammed-earth dome can withstand an enormous amount of weight.

Shingled Dome with Extended Eaves

We have yet to try it on a full-scale dome project, but we've collected loads of free asphalt shingles and hammered them into the mud-plastered surface of the Honey House dome. They are easy to anchor, and stagger to conform to the curve of the roof. They can be installed very thickly with an overlap of one inch (2.5 cm) or less creating a thick, thatched appearance. Earthbag domes can carry a substantial load. With shingles set this thick they could last a lifetime, and we may have developed safe and effective ways to recycle them by the time they need to be replaced.

Since asphalt shingles do not breathe, a cupola might be in order. This may be built in such a way that any moisture vapor that builds up under the shingles can travel upward, perhaps along channels sculpted into a base coat of earthen plaster, and vent out the cupola. The eaves can be extended to rest on a wrap-around portico of earthbag arches or a post and beam porch.

Wood shakes and shingles are a natural, breathable alternative to asphalt or fiberglass shingles, and can be applied directly over well-executed bag work without any base coat of plaster. In essence, any kind of roof tile from slate to terra cotta to slab stone can be easily supported by a properly constructed earthbag dome. Experiment, explore, heighten, and discover! (Fig. 13.8).

13.8: *Shingles applied directly over an earthbag dome with built-in extended eaves to protect walls.*

Insulated Light Wood Frame

People living in high rainfall areas, heavy snow country, or areas with abundant access to wood products, can secure light, wood frame rafters to built-in extended eaves. We recommend light wood frame as an alternative to larger dimensional lumber. Not because the dome can't take the weight, but to reduce timber consumption (Fig. 13.9).

Exterior wood-frame roof systems can be insulated, sheathed with wood, and covered with any kind of roofing material, like metal, wood, asphalt shingles, or Eco-shingles. A wood frame can be built right over the bag work of the dome without any mud plaster (Fig. 13.10).

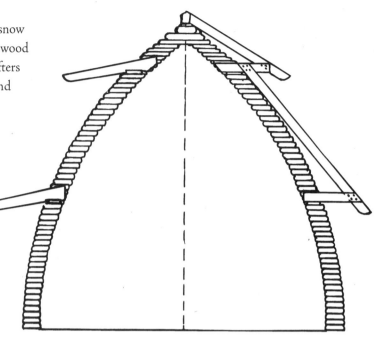

The "Lighthouse" style extended eaves only

"Japanese bird cage" style light wood frame attached to extended eaves

13.9: *Earthbag dome with a light, wood-frame roof attached to built-in eaves.*

The Lighthouse style, extended eaves only

Japanese bird cage style, light wood frame attached to extended eaves

External Straw-Bale-Insulated Roof

Stacking straw bales on top of long, beefy extended eaves can create another version of a living roof. The eaves would need to be sheathed, and some sort of waterproof membrane, like sturdy EPDM heavy plastic pond liner, cut into large sections and overlapped like shingles over the entire dome. The bales can then be stacked and wedged at an angle with bale flakes, cob, clay/straw blocks, etc. An earthen mud mortar slathered in between each course of bales will help to stick them together. Wire cable cinched tight at the center of every two or three courses is another way to secure them.

13.10: *Split-screen image of two styles of wood roof over an earthbag dome.*

In a dry climate, the cap can be plugged with sculpted cob and lime plaster. Straw bales last many years exposed in the desert as long as they are kept up off the ground. In a moister climate, the bales can be encouraged to grow into a living roof. Planting tough grasses with migrating root systems (like Bermuda), or sturdy vines (Virginia creeper, ivy, honeysuckle, etc) onto the surface of the bales and into the mud material packed between them would be fun to try. The initial cost is minimal; if the bales fail over time, they can be replaced with more bales or a shingled roof.

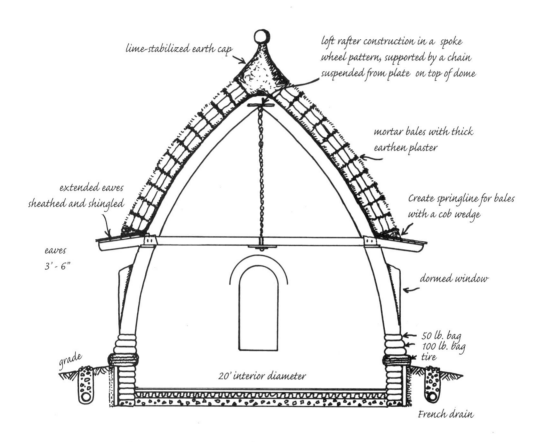

lime-stabilized earth cap

loft rafter construction in a spoke wheel pattern, supported by a chain suspended from plate on top of dome

mortar bales with thick earthen plaster

extended eaves sheathed and shingled

Create springline for bales with a cob wedge

eaves 3' - 6"

dormed window

grade

50 lb. bag
100 lb. bag
tire

20' interior diameter

French drain

13.11: *Scale model of a 20-foot (6 m) diameter earthbag dome with external straw-bale-insulated roof.*

CHAPTER 14

Exterior Plasters

Earthbag walls can be likened to bare bones await-ing a fleshy coat of protective plaster. In this chapter, we will look at earthen plasters, lime over an earthen plaster, and lastly address the use of cement. Let's approach the subject of earthen plasters by exam-ining the basics, as we have done previously with earthen building.

Earthen Plaster

Earthen and lime plasters absorb and transpire mois-ture out through the walls, helping to regulate internal humidity. Properly mixed and applied, earthen plas-ters are mold, vermin, and UV resistant, long-lasting, durable, and above all, beautiful. Earthen plasters are a joy to work with and live within. They are naturally soothing to the senses and have mild detoxifying properties.

JEANETTE POWERS

14.1: *Earthbag walls with an earthen plaster coat, in mid-winter.*

14.2: *Modern day enharradoras.*

We live in a dry climate where a clay-rich earthen plaster with lots of long and chopped straw holds up well on exterior walls, with replastering touch-ups every few years. A clay-rich earthen plaster will stick to just about anything, eliminating the need for chicken wire lath. Earth plasters high in clay are more weather resistant than low-clay high-sand plasters that tend to erode very easily. When a clay-rich plaster gets wet, the clay swells slightly, inhibiting water from penetrating too deep.

14.3: *Softening clay lumps in a mud soaking pit.*

In climates where rainfall is low, the plaster quickly dries out and remains stable. Our original earthbag walls, built in 1994, still have the same primitive earth plaster covering on the vertical surfaces, and a one-time replastered topcoat on the horizontal surface.

The Spanish word for an adobe plasterer is *enharradora,* and the women traditionally did the plastering. *Enharradoras* in the Southwest desert say that it takes ten years to wash away a quarter-inch (0.625 cm) of earthen plaster on an exposed vertical wall surface (Fig. 14.2).

The Big Three: Clay, Sand, and Fiber

Clay, sand, and fiber are the three magical ingredients in recipes for earth plaster, cob, and poured adobe floors. All we do is adjust the ratios, fine tune the texture, and apply the mix in thicker or thinner layers. Clay now performs the starring role. Sand and fiber are relegated to supporting roles, yet still play a prominent part.

Clay (Fig. 14.3)

If earth plaster were a drug, clay would be listed on the label as the active ingredient. For earthen plasters, clay is the essential element that holds sand and fiber together and adheres to a wall. Do a jar test to determine the ratios of a proposed plaster soil candidate. Play with the stuff. It should feel sticky, plastic, moldable, and pleasurable to handle. Let thy hands be thine guide (refer to the section The Dirt, in Chapter 2, for more exhaustive explanations on the merits of clay).

Sand

By playing the role of supporting actor, sand provides abrasion resistance, compressive strength, and reduced shrinkage. Well-graded coarse sand is the natural optimum for plasters, sifted through a fine screen of up to half-inch (1.25 cm) mesh depending on its intended use.

Fiber

The fiber we use most commonly in earthen plaster is straw. Straw is derived from the stalk of grains such as oat, wheat, barley, rice, etc. Short, chopped straw is most often used for plaster, but the longer the straw, the coarser and stronger the mix. Straw provides tensile strength, much like chicken wire does for cement stucco. Straw performs like a mesh that weaves the surface together into a monolithic blanket. It helps resist erosion by exposing a matrix of little diversion routes that distribute rainwater evenly down the surface of a wall or the roof of a dome. Exterior plasters with an ample amount of long straw provide the most resistance to erosion.

We have seen how a beautiful earthen plaster prepared with short straw disintegrated into a fluffy golden bird's nest around the bottom perimeter walls after being pummeled by a violent thundershower. Exterior plaster made with primarily long straw as its fiber stays put. Other fibers we may use, in addition to long straw (depending on location, availability, and desired effect), include sun-bleached grass clippings, sawdust, slurried cow or horse manure, shredded paper, fibrous tree bark, hemp, sisal, coconut husk, cattail catkins, coarse, short animal hair, etc. A variety of fibers in combination gives the desired textured finish while enhancing tensile strength and resistance to erosion (Fig. 14.4).

Plastering Equipment

We started out making and applying earthen plaster using the simplest equipment, our feet and hands. Gradually, we progressed to an assortment of tools and even some machinery. Start out simply and add tools as you become more proficient. The right tool can save time and create desired effects that may not be achievable without them. Pool trowels are fun to use and a small cement mixer can really be helpful for individuals working alone, or to keep up with a large crew.

Test Batches

Get to know the earth by playing with it. When we travel to conduct workshops, we have to start from scratch to discover the merits of a new batch of soil. The key is in experimentation. The fun-d-mental principle is this: adapt the ratios of the three magical ingredients until the plaster behaves the way you want it to. As a general rule of thumb, 30 percent clay to 70 percent well-graded sand is optimal with enough straw to give it body and eliminate cracking. Straw accounts for about 20 percent-50percent by volume according to personal appeal and the behavior of the mix.

14.4: A variety of well-graded fibers for making plaster: clockwise from upper left; paper cellulose, chopped straw, long straw (on ground), sun-bleached grass clippings, and horse manure.

COMMON EQUIPMENT FOR MIXING AND PREPARING PLASTER

- Feet and hands
- Shovel
- Concrete or garden hoe
- Wheelbarrow
- Tarpaulins
- Straw bales (to make a ring for a mixing pit)
- Large screen (one-quarter to half-inch [0.625-1.25 cm] hardware cloth over a rigid frame)
- Semi-fine screen (kitchen wire mesh colander)
- Hose with spray gun
- Chipper/shredder or machete to chop straw
- Mortar or cement mixer (optional)

Application Tools
- Your hands
- Buckets and cans
- Fat paint brush, or garden sprayer, or hose with spray gun (for dampening cured plaster or cob)
- Assorted trowels - steel pool trowel, square-edge margin trowel, pointed trowel (for hard to reach corners), wood float, sponge float (or thick cellulose sponge)

- Dimpler (for texturing surface to provide key-in for successive plaster coats)
- Hawk (optional tool used by "the pros")

Other Equipment Handy for Making the Job Easier
- Plastic and paper (for protecting windows, doors, floors, and wood or metal trim)
- Tape — blue (quick release), masking, and duct (for attaching the plastic or paper to windows, doors, and trim)
- Ladders
- Scaffolding or planks over saw horses or straw bales

Optional Equipment
- Air compressor
- Drywall texture gun and/or plaster sprayer

Some or all of these tools and equipment can be used for any type of plaster project, be it earthen, lime, cement, or cob.

Use a section of the wall as a sample board. The area should be at least one-square-foot (0.1 sq. meter) in size for each test batch. Let it dry. If the plaster sample shrinks or cracks a lot, add more sand or straw or both. A few small hairline cracks are all right, as long as the plaster adheres to the wall without trying to curl away. The following layer will fill in these small cracks. If the plaster dries powdery and weak, the clay could be too expansive or the earth have too high a silt content. If clay is scarce or the soil in general is of a poor quality, we resort to adding a binder of cooked flour paste (see recipe under "Additives for Fat Plaster" in Chapter 15).

How to Prepare Exterior Earthen Plaster

After years of stomping mud with our feet, we bought a used cement mixer for mixing earthen plaster and cob. Small new and older used cement mixers are cheaper and easier to come by than mortar mixers. They can also handle a coarser earthen mix (less screening!) with an abundance of long straw.

Sift all the earthen ingredients through a quarter-inch (0.625 cm) screen or, to avoid having to pulverize and screen large, dry clay lumps, presoak the soil in a pit close to the mixing area. Allow the soil to percolate overnight or until the clay has softened. If using a cement mixer, start with about two gallons (7.5 liters) of water in the mixer. Throw in several fist-loads of long straw. Shovel in the pre-softened dirt until the consistency in the mixer is that of thick cake batter (adding the straw first helps keep the mud from splashing out). If you are using earth of two different qualities, like a clayey soil and reject sand, add the proper ratio determined from test batches (for example, seven shovels of reject to three shovels of clay, etc.). Add more water, and keep adding straw (and any other desired fibers) and soil until the mix starts to ball up. The soil will sometimes glob up and stick to the bottom of the mixer. If this happens, stop the mixer, pull the glob off the bottom and sides, shove a few handfuls of straw into the back, and give it another spin (Fig. 14.5).

14.5: *Adding long straw to a batch of earthen plaster.*

Foot mixing can be done in a shallow pit in the ground lined with a tarp, old bathtub, water trough, or a ring of straw bales with a tarp lining them (Fig. 14.6).

If you are dealing with dry soil, screen all the soil first to break up the clumps (tampers work great for pulverizing clay lumps). Fill the pit or container with

14.6: *Presoak the soil in a pit, and add handfuls of straw while stomping around in the muck.*

an appropriate amount of water and broadcast the screened soil into it with a shovel. Screening helps dissolve the soil quicker and is easier on the feet. Layer the mix with straw, and get in there and stomp around.

Procedure for Hand Application Directly Over Earthbags

Lath Coat

Jab fingers full of fiber-rich, firm yet sticky plaster in between the rows of bags, like chinking in between a log wall. Be generous. Apply the lath coat thickly enough that the plaster extends a little past the surface of the bags. Leave it rough with plenty of finger poke holes to provide a mechanical bond (Fig. 14.7a).

After this first lath coat has set up firm, but is still green, slap on three-quarters- to one-inch (2-2.5 thick cm) patties of the same mix, perhaps with additional long straw. Fill in any low spots with more patties (Fig. 14.7b). Trowel smooth as a finish coat or

14.7a (above): *Chinking coat*
14.7b (below): *Patty smacking*

leave textured as a key-in for subsequent earthen or lime plaster finishes (Fig. 14.7c & 14.7d). Throughout the application process, keep your application hand and trowel wet and slippery to encourage the mud to stick to the wall instead of to your hands and tools.

14.7c (above): *Troweling patties*
14.7d (below): *Finger poking for key-in.*

NOTE: If applying the mud to burlap bags, pre-moisten the wall. It is not necessary to pre-moisten the poly bags.

Mechanical Bond

The *mechanical bond* refers to the physical connection of plaster to the surface of a wall. The surface of a preceding plaster layer needs to be highly textured in order to provide a "key" to lock in the following coats. A well-keyed surface is another term for mechanical bond. Chicken wire provides the mechanical bond for conventional cement stucco. Roughing up the surface of an earthen plaster provides a mechanical bond for subsequent layers of earthen or lime plaster.

The base coat is the thickest layer of plaster. The base coat is also often the *fill coat*. A fill coat is traditionally used to build up low spots as well as begin the foundation for sculptural relief work. It is the most highly textured layer of plaster. We like to use our fingers to poke the surface full of dimples. A *scratch board* (basically, a piece of wood with a bunch of nails poking through one side) is designed to prick the surface full of lots of little holes. A stiff rake or broom can also be used, but care should be taken as these tools have a tendency to drag the straw out of the wall (Fig. 14.8).

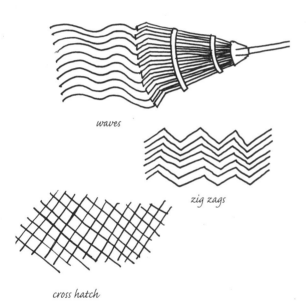

waves

zig zags

cross hatch

14.8: *Additional methods for creating a mechanical bond in fiber-free plaster.*

Misting

After a plaster has fully dried, refresh the surface with water, either sprayed on or applied with a fat paintbrush. Dampening the surface reactivates the clay and aids in bonding the two layers together. Allow the water to soak in for a few minutes before applying the next layer of plaster. The surface should be moist but not glistening.

Mortar Coat

Over the years we have made it part of our repertoire to apply what we call a *mortar coat* to the surface of a cured plaster just before applying the next coat. The mortar coat has the consistency of soft cream cheese. It is a strawless version of the same plaster mix we are using. It acts like glue to make the second coat stick much more easily. Mortar or "smear coats" (as the kids like to call it) are used mostly for cementing a thick layer of plaster that is often slapped on like a patty and then troweled or smoothed over by hand.

Application Over Chicken Wire

The underside of the arches and any window openings that were lined with wire mesh receive a primer coat of strawless plaster so the plaster will slip easily through the mesh. If whole walls have been wired, we may substitute sawdust shavings for straw that is smaller than the diameter of the chicken wire mesh being used. The sawdust reduces cracking and adds a fair amount of tensile strength.

Sawdust tends to be more sensitive to mold, so we usually add about a cup of Borax to an average size wheelbarrow load of plaster during mixing. The Borax is a highly alkaline mineral that naturally inhibits mildew (see "Additives" in Chapter 15). We then switch back to a straw fortified plaster mix for the second coat, as straw is more weather resistant.

Mechanical Application

Spray Guns

Another strategy is to "shoot" an earthen plaster with a *plaster sprayer* making sure the earth and straw is screened fine enough to shoot through the nozzle of the sprayer. In addition to fine chopped straw, shredded paper and slurried cow or horse manure help to round out the fiber mix. The addition of manure to your mix, while it might sound offensive, gives the added benefit of increasing the stickiness of the mix and appears to add a degree of water resistance to a plaster when dry. The short fiber required for use with a spray gun makes the plaster less resistant to weather erosion, but works fine as a base coat for either a long straw-rich earthen finish coat or as a key-in coat for lime plaster (Fig. 14.9 & 14.10).

(Check the Resources Guide for contact information concerning the plaster sprayer).

14.9: *Shooting mud onto an earthbag wall with a plaster sprayer.*

14.10: *A spray-on application finish preserves the look of rows of bags or coiled tubes. Photo Credit: Mara Cranic*

Uses of Cob with Earthbags

We employ cob for a myriad of purposes. We use it as a sculptural medium for fleshing out dormered windows, gutter systems, built-in furniture, and floors, and as a base layer over a dome designed for an earthen or lime plaster finish. We refer to cob as the duct tape of natural building due to its invaluable variety of uses (Fig. 14.11).

In the Southwestern US, cob is also called *sculptural*, *monolithic*, or *coursed adobe*. The emphasis is on the mix being a stiff soil with a clay content of about 30 percent, with the balance made up of well-graded, coarse sand and as much long straw as it will hold and still be sticky. The reject sand we get from our local gravel yard forms the basis of our cob mix, with the addition of a few extra shovels of high-clay-content soil and gobs of long straw. A well-graded

coarse sand is optimum for an earthen building soil, but for a sculptural medium, we can use a wider variety of soils, as long as the clay content is high enough and of a stable quality. We like to make cob in a cement mixer, adding copious amounts of straw. The mixer will literally spit out loaves when the mix is done. The cob loaves should feel firm but not dry (Fig. 14.12).

To provide a key-in for relief work, drive nails into the wall so that they protrude at opposing angles. Installing extra-wide chicken wire cradles around the fan bags surrounding an arched opening

14.11: *Sculpting the secondstory dormers around the arch forms of the Honey House.*

provides built-in pockets to fill with cob for building out drip edges, or for interior relief work. Leave the surface of the cob rough to receive additional plaster layers.

We first learned about cob in a book entitled *The Bread Ovens of Quebec* that examines the historical construction of clay ovens in Quebec, Canada, since the mid-17th century (Fig. 14.13).

14.12: *To mix cob loaves stiffer than the mixer, we throw a chunk of the mix into a wheelbarrow lined with straw and knead it like bread, jabbing the center of the loaves with our thumbs.*

14.13: *Making our first cob oven was the inspiration for building the Honey House earthbag dome.*

Lime Plaster

Many people live where the annual rainfall is higher than in the desert Southwest. You can use an earthen plaster as a base coat, but for more serious protection from erosion, a lime plaster is a natural substance that can be applied over an earthen plaster. Lime plaster will set over time to a hard, erosion-resistant finish, like a natural Goretex wall covering. It allows transpiration of water vapor while reducing water penetration. This, in turn, preserves the structural integrity of an earthen wall. Plasters and mortars made from fired limestone have been the traditional renders used to protect natural wall systems for thousands of years.

There are a few different types of lime that are used to ultimately achieve the same effect. The two most common types are determined by the amount of calcium present in the parent limestone rock. A high-calcium lime usually contains over 94 percent calcium, while a dolomitic lime contains about 30 percent magnesium with the remainder made up of calcium. A primary difference between these two types of lime is in how long it takes them to harden on the walls, or more specifically, their individual rates of *carbonation*.

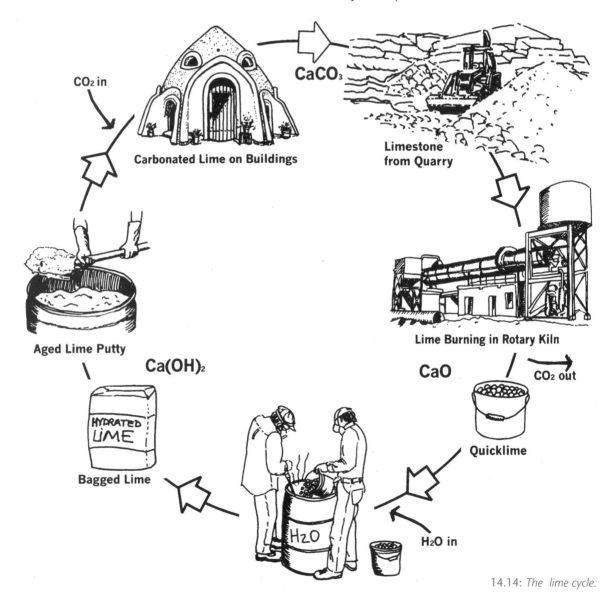

CO_2 in

Carbonated Lime on Buildings

$CaCO_3$

Limestone from Quarry

Aged Lime Putty

$Ca(OH)_2$

HYDRATED LIME

Bagged Lime

Lime Burning in Rotary Kiln

CaO

CO_2 out

Quicklime

H_2O

H_2O in

14.14: *The lime cycle.*

The Lime Cycle (Fig. 14.14)

High-calcium limestone, in its native rock state, has the chemical composition $CaCO_3$, calcium carbonate. CO_2 (carbon dioxide) is driven out of the limestone in the process of firing the rock creating CaO, calcium oxide (also called "quicklime" or "lump lime"). Dolomitic limestone, on the other hand, has the formula $CaMgCO_3$, and when fired ends up as CaMgO, losing a CO_2 molecule but retaining both the calcium and magnesium. When either of these fired limestones, or quicklime, is recombined with water (a process referred to as *slaking*), an exothermic reaction takes place that converts the calcium oxide into calcium hydroxide (Ca $(OH)_2$), or *hydrated lime*. The main difference is that the high calcium variety reacts several times more quickly than the high magnesium sample. When this calcium hydroxide, commonly known as lime putty, is mixed with sand and applied onto a wall surface as a plaster, another chemical reaction takes place that essentially reintroduces carbon dioxide (CO_2) back into the lime; over time it will harden and revert back to limestone. This process is known as carbonation and, just as the two types of limestone react differently when slaked, they also recarbonate at different rates, the high calcium variety being the fastest.

Unfortunately, quicklime is not readily available to the owner/builder without a lot of investigative searching. The hazards associated with the slaking of quicklime also act as a deterrent to the first time do-it-yourselfer. As a result, the most readily available building lime in the US is the pre-bagged variety, commonly referred to as *Type S-Hydrated Lime*.

Type S-Hydrated Lime

Type S lime is mined from magnesium-rich dolomite limestone that has been fired to produce quicklime, ground into a powder, and then hydrated under pressure in a controlled manufacturing process with just enough water to thoroughly react with the quicklime without saturating it. It is sold bagged in a powder form. Type S lime should be purchased in as fresh a state as possible.

The problem with using Type S-Hydrated Lime if the bag is not fresh is that the carbonation process may have already begun. If the lime is exposed to moisture in the air, over time it will recarbonate in the bag, reverting back into calcium carbonate (limestone), losing all of its binding property. Slaking the lime and storing it in a putty state until you are ready to use it will stop the carbonation process by inhibiting exposure to the air.

Aged lime putty should be thick enough to sit up on a shovel. If the putty is too runny it will have little binding power. Although successful plaster can be achieved using Type S-Hydrated Lime, our

SLAKING TYPE S LIME PUTTY

Here is what we consider a simple method for making lime putty from Type S lime hydrate: fill a plastic garbage can with 16 gallons (60 liters) of clean water. Shovel in two 50-pound (22.2 kg) bags of lime powder. If the lime is at all lumpy, place a quarter-inch (0.625 cm) screen over the top of the can and push the lime through it. Stir any exposed lime down into the water. Add as much additional lime as the water will hold until the mix is the consistency of thick sour cream or thicker. The whole mix can be stirred in the can with a long-handled mixing paddle attached to a heavy-duty drill motor. Once the mix is thick, add about an inch (2.5 cm) more of water on top to protect the putty from exposure to air, and seal it with a tight fitting lid. Protect it from freezing, as it will curd up like frozen cream cheese.

14.15: *Slaking lime putty.*

personal experience with both Type S and home slaked high-calcium lime that has continued to age into its third year has shown that the high-calcium variety is far superior to Type S. The binding properties and accelerated carbonization rate of this high quality high-calcium lime is superior to any lime made from hydrated lime we have tried.

Mixing Lime Plaster

When mixing lime with sand to create a plaster, a typical mix contains two and one-half to three parts clean, well-graded sand with one part firm lime putty. Typically, a base coat lime plaster uses a coarse washed "concrete grade" sand followed by finer coats using finer "mortar grade" sand. As with all of our recommendations, experimentation is a critical part of learning how the medium you are working with responds. Make some test patches using different ratios of sand to lime putty and allow your test patches to fully cure before picking the mix best suited to your particular application.

Like earthen plaster, lime plasters can also benefit from the addition of compatible finely chopped fiber such as slurried cow manure, coarse animal hair, straw chaff, or finely chopped sisal. Usually, fibers that are high in either silica content or protein (animal hair) are suitable. Lime plasters and washes can be tinted with colored oxide pigments that are compatible with lime (at a ratio of up to 10 percent diluted pigment to putty). A little extra time spent doing these tests more than pays for itself in the long run. It also adds to your understanding of what it is you are working with.

Application

Lime plasters can be applied either by hand or by mechanical means, and adhere directly to an earthen substrate when applied with some force. *Harling* (casting or spraying the plaster onto the surface) provides a better bond than simply troweling it on, particularly for the initial coat. Avoid overworking lime with a steel trowel, as the metal tends to lift the lime particles to the surface and promote cracking. Lime worked with a wood trowel, however, maintains microscopic pore spaces that readily accept another lime plaster coat, or lime washes. A few thin layers are more effective than one thick layer. Successive coats may be applied by hand, or troweled after the previous coat has set up to the point that it can be dented with a fingernail, but not with a finger. Prior to applying the next coat of plaster, always pre-moisten the wall. Allow the water to soak in, and then apply the next coat (Fig. 14.16a, b & c).

14.16a: *Hand application*   14.16b: *Trowel application*   14.16c: *Texture gun spray-on application.*

Curing

Lime needs time to cure suitably to ensure proper carbonization. Keeping the walls damp, but not soggy, during application and for at least one week afterwards, is necessary for complete curing to occur. The process of wetting and partial drying transports carbon into the matrix of the lime plaster, enabling the carbonization process.

Protection

Lime plaster needs frost-free moist curing conditions. Lime plaster is best applied at least two weeks before any possibility of frost. External plastic tarps are one way to seal in a moist environment in a dry climate or protect from too much rain or possible frost in a wet/cold climate.

Safety Precautions

One should use care when dealing with lime. It is a very caustic material that can cause severe burns on skin, in the eyes, or mucous membranes. Always protect yourself by wearing safety glasses, rubber gloves, long pants, long-sleeved shirts, and sturdy shoes when handling lime. Keep some clean water mixed half-and-half with vinegar on hand while working with lime and wash off any errant splashes. The acid in the vinegar counteracts the alkalinity of the lime. Don't let these warnings deter you from working with this remarkable medium.

Lime Water

When pre-moistening an earthen wall surface prior to applying a lime render, it is advisable to use water that has been treated with a small percentage of lime. This allows the carbonation process to occur in the outer clay surface of the wall, creating a chemical bond between the earthen substrate and the lime plaster.

As lime putty settles, the clear water on the top is known as limewater. A clear layer of calcite crystals forms on the surface. Simply stated, limewater is a

saturated solution of calcium hydroxide in water. The easiest way to obtain limewater for prepping wall surfaces is to use the water that sits directly on top of lime putty. Nothing more needs to be added or done to it. Remember to add a little water to the top of your lime putty to replace what is used. If your putty has no water on top (uh, oh!) a few tablespoons of putty mixed with a gallon (3.75 lires) of clean water is sufficient to make limewater.

Lime Wash

Lime wash, sometimes known as *milk of lime* (according to Holmes and Wingate in their book, *Building With Lime*), is "a free flowing suspension of hydrated lime in water in such proportions as to resemble milk in appearance." It's a simple form of paint prepared from lime. It can be used alone or tinted with mineral pigments to create pastel colors. Lime wash is used as a periodic maintenance for lime plasters as it fills small cracks and has the amazing ability to self-heal. As cracks develop, the lime wash creates crystal growth to seal the blemish. Lime wash is most often used over lime plasters to finish the plaster and make it more resistant to cracking and weathering (Fig. 14.17).

To make a lime wash, combine about one gallon (3.75 liters) of stiff lime putty to about four gallons (15 liters) of water. Mix to the consistency of milk. If

14.17: One-hundred-year-old adobe house protected with a coarse coat of lime plaster.

it is too thin, add a little more putty. If too thick, add more water until the desired consistency is achieved. If it is too thick, the wash may show many fine cracks when dry, known as "crazing." It is usually best to mix on the thin side and apply several coats to build it up to the desired thickness. A thicker version of lime wash is made by adding fine silica sand (50-70 grit) at a ratio of 50:50 sand to lime putty.

14.18: Applying cement-lime stucco over chicken wire on the Sand Castle on Rum Cay, Bahamas. Photo Credit: Steve Kemble and Carol Escott

The Future of Lime

Lime plaster has been used successfully for centuries and is still as available as Ready-Mix concrete in Great Britain. It is currently seeing a resurgence of use here in the US as its merits and benefits become more apparent. Perhaps as demand grows, so will a market for aged lime putty slaked from high-calcium quicklime. (For more in-depth information on lime, refer to the references listed in the Resource Guide at the end of this book).

Cement Plaster (Fig. 14.18)

About 75 to 100 years ago, cement became available to the general public. It was supposed to be the end-all to the desire for a quick and hard setting, low maintenance plaster and mortar that would replace the softer, slower-setting traditional lime mortars and plasters. We flocked to it like moths to a flame, only to be burned in the long run. Many historic missions in the Southwest have been brought to their knees due to the substitution of cement for traditional lime-based renders.

As we mentioned in Chapter 4, the problem with cement stucco over earthen substrates is that cement is impermeable. That is, it does not allow water vapor to transpire through it. The funny thing is, it wicks water readily thus accumulating water in the underlying layers. Water passes through cement, but water *vapor* does not. This is the main reason that so many adobe missions in the Southwest that had their traditional lime plasters replaced with cement stucco in the last century are now showing signs of imminent failure (Fig. 14.19).

We are obviously not big proponents of using cement. As noted earlier in this book, cement manufacturing contributes a significant amount of pollution to our environment. But situations arise in this big world that sometimes require us to eat a little crow

14.19: *Even cement plaster over stabilized adobe has continued to crack on the walls of this high-end Southwest-style home.*

and admit there are some very good uses for a material that we normally wouldn't be caught dead working with. There are some situations where cement plaster is not only a good plaster, but also the best choice, like building in places where clay simply does not exist.

While we were in the Bahamas helping our friends with their earthbag project, we were often vexed with the unavailability of some construction materials. We take for granted a trip to the hardware store for something we don't have, but when the nearest store is over 200 miles (320 km) away by boat, it's necessary to simply make-do. What they did have an abundance of was cement and cement-based products. Even though the old ruins on the island contained lime mortars and plasters that had undoubtedly been produced locally, these remote locations had also fallen prey to the inscrutable promotion of cement in the 20th century.

If you live in an area that is rarely or never subjected to freezing weather (like the Bahamas), it is probably all right to use cement-based stucco over earthbags. Cement-based plasters are most suitable over earthbags filled with coarse, sandy, well-draining soils, as sandy soils are less apt to harbor moisture and therefore to remain stable in wet or freezing conditions.

Chicken wire, stucco mesh, or extruded plastic mesh can be used in conjunction with cement plaster. The mesh provides tensile strength for the cement, which is likely to separate from the surface of the earthbags without it.

Stucco Wire/Plastic Mesh Installation

The simplest way to install stucco mesh is to tack it up along its vertical edge onto the wall and unroll it across the whole surface of the wall. Stretch it tight as you go and tack with 2-2½-inch (5 cm-6.25 cm) galvanized roofing nails into the surface of the bags. A good quality, cured rammed-earth soil will hold nails — even bend them in some places. If, however, the soil is soft, then you will have to rely on tie wires that you hopefully installed during wall construction, based on the results from your preliminary soil tests (Fig. 14.20).

We go ahead and stretch the wire over the entire wall, including any windows and doors, then cinch it tight and tack it down. The mesh should be snug, but not so tight that the wire is sucked too deep in between the bags. Try to keep it flush with the protruding surface of the bags over the whole wall area. Then twist the wires tight to take up any slack. For esthetic appeal, you may want to soften any inside corners by rounding them out with the wire mesh.

Since our main focus is on the use of natural materials, and less cement, we will refrain here from describing how to mix and apply cement onto an earthbag building. The general process of mixing and applying cement is covered very extensively in trade

14.20: *Carol Escott cinches up chicken wire with a combination of tie wires and nails. Photo Credit: Sustainable Systems Support.*

journals and books by cement proponents, who know and enjoy the medium more than we.

Gauging Cement with Lime

If an earthbag structure is located in a wet area that is subject to freezing in the winter, cement stucco can be used if lime is added to the mix. While the addition of cement to lime will adversely affect lime's ability to fully carbonate, adding up to 50 percent lime to cement (by volume) increases cement's ability to transpire moisture, while maintaining the binding properties inherent in cement. In this situation, we still recommend filling the bags with a coarse, sandy, well-draining soil to inhibit moisture buildup in the walls, and decrease cracking on the surface of the plaster.

Conclusion

The main thing to remember is that we are working with natural materials that respond to the environment and its subtle and not so subtle changes. Earthen plasters swell and contract in response to changes in humidity, temperature, and even seasonally. What we try to provide is a finish exterior coat that allows the walls the freedom to adapt to the changing environment while maintaining their structural integrity, that provide protection from the ravages of inclement weather, and that offer, as well, a source of beauty and simplicity.

CHAPTER 15

Interior Plasters

In the summer of 2002 we were hired to finish the interior of a load-bearing earthbag building designed by the US Interior Department, Bureau of Land Management. This 750 square foot (72.5 sq. m) Pueblo style Ranger contact station sits along the shore of the San Juan River at the Sand Island boat ramp in southeastern Utah. We had been contracted the year before to train a crew and assist with the earthbag construction. Our tax dollars at work! (Fig 15.2).

Although we give them full credit for the exterior finish of Dryvit over foam over plywood (despite our suggestion for more compatible systems), they gave us creative license to finish the interior with locally

15.1 (above): *Moon House ruin, Utah. Original plaster, circa 1200 AD.*

15.2 (left): *We were thrilled with the opportunity to install a totally-natural earthen wall finish in a government building open to the public.*

harvested wild clays from our public lands and a sandy soil from a nearby gravel yard.

Our focus for interior plasters is primarily earthen. We choose this medium for its availability, ease of use, familiarity, and because of its compatibility with the earthbag building system. Where the components for an earthen plaster are unavailable, cement, lime, and gypsum are viable options. (For more information on cement and lime applications, see Chapter 14).

Fat Plaster

Because our choice of plaster for this project had to live up to the expectations of a public use facility, we were challenged to develop a new standard for a durable, mold resistant, easy to apply, earthen plaster. Hence, *Fat plaster* was born. Fat plaster is just that. It's a very thick coat of plaster that goes on in two stages. Fat plaster is strong, durable, and easy to apply. If the color of the earth is attractive, the plaster can be troweled smooth and left as the finished wall, or it can be lightly textured to receive subsequent finish coats of a more refined plaster, *alis* (clay paint) or milk paint. Fat plaster reduces prep time by being able to use coarser materials. This means less time is spent screening the materials. Fat plaster can be customized to perform as an effective plaster method for applying plaster onto earthbags with or without the installation of chicken wire lath.

The Key to Successful Fat Plaster is in the Variety of Fiber Sizes

Just as an earth-building soil or earthen plaster benefits from the inclusion of well-graded sand, we have found the same benefit to apply to the fiber in a base-coat fat plaster. Although many options for fiber exist, we selected three varieties that have proven, so far, to suit our purposes the best. The three varieties of fiber we settled on are: chopped straw from chaff up to 1½-inches (3.75 cm) long, sun bleached grass clippings up to 1½ inches long, and paper cellulose (like the kind used as blown-in cellulose insulation).

Straw provides a little bulk and adds tensile strength throughout the matrix. The fine grass clippings also add tensile strength, while making the plaster more malleable and easy to sculpt. The paper cellulose is the crème-de-la-crème, making the plaster super creamy, while providing tensile strength on a microscopic level. The combination of three distinct sizes of well-graded fiber mixed with a typical plaster soil of (approximately) 25 percent clay and 75 percent well-graded sand has produced a strong crack-resistant plaster that is fun and easy to apply.

Additives for Fat Plaster

In addition to the fiber, we also add a mold inhibitor. Although interior environments are safe from erosion due to weather, they still generate a considerable amount of internal moisture from showers, cooking, houseplants, and, specific to the Southwest, evaporative cooling systems, quaintly, yet accurately referred to as "swamp coolers."

Borax

Organic fibers are at risk for the ravages of mold as the result of trapped moisture. We add a small portion of borax (the type sold in stores as a laundry booster) to our plaster. About one cup of borax to every four-cubic-foot wheelbarrow seems to work well. To prepare borax, screen the powder through a fine kitchen colander screen and mix this screened powder with one cup of water. Add this mixture to your plaster water.

Cooked Flour Paste

The other additive we have become addicted to is cooked flour paste. Cooked flour paste adds additional strength and prevents the dried plaster from dusting off the wall. It also provides additional stickiness during the application process and can be used to bolster the binding properties of a plaster lacking in clay, or clay of poor binding strength.

Flour paste also adds a sharp, sand-papery surface to a wall when the still-green plaster is sponged or rubbed with rubber gloves, making an excellent, toothy surface for a later application of an alis or milk paint. The surface can be more severely raked or dimpled for a troweled-on finish plaster, or hand troweled smooth and left finished as is. In general, we add anywhere from 1½ to 3-quarts (1.4-2.8 liters) of cooked flour paste to a four-cubic-foot wheelbarrow load. Every soil is different, of course, so you'll need to experiment with the ratios.

Aside from the borax and cooked flour paste, we don't use any other additives for making fat plaster. There are many more options for plaster recipes with all kinds of additives. This is just our personal preference and we have found them to be effective and enjoyable plaster mixes for using directly over earthbags.

Fat Plaster Recipe (Fig. 15.3)

- 3-4 gallons (11.3-15 liters) of clean water (total amount needed)
- 2 #10 cans of chopped straw
- 2 #10 cans of dried grass clippings
- 4 #10 cans of paper cellulose
- 6 shovels of clay-rich soil, screened to ¼ or ½ inch (0.625 or 1.25 cm)
- 18 shovels sandy soil (screened to ¼ or ½ inch [0.625 or 1.25 cm])
- 1.5-3 quarts (1.4-2.8 liters) of cooked flour paste
- 1 cup of borax dissolved in 1 cup of water

This recipe will make an average four-cubic-foot wheelbarrow load. Everyone tends to develop his or her own style of mixing. Usually it's best to start with a portion of the water (say one-half to two-thirds of the total amount). We add the diluted borax and flour paste in the beginning to thoroughly amalgamate the borax, flour paste, and water. We add the clay-rich soil next to ensure that it is completely saturated. If after adding the clay the mix becomes too thick, add enough water to keep it soupy. Then we add the fiber and sand, using a little more or less sand to achieve the desired consistency. We strive for a consistency that is firm enough to mold into a ball, yet pliable enough to sculpt with a trowel.

RECIPE FOR FOUR GALLONS (15 LITERS) OF COOKED FLOUR PASTE

In a large 20-quart (18.75 liter) canning pot, bring 3 gallons (11.25 liters) of water to a boil. In another container, add 10 cups white wheat flour to 1 gallon (3.75 liters) of cold water. Whisk this mixture to a creamy consistency. When the 3 gallons of water come to a roaring boil, quickly pour the creamy flour batter into it while whisking thoroughly. It will turn from an opaque white to a thickened, translucent, jelly-like consistency. We like to cook it a few minutes longer, stirring the bottom with a long-handled, heatproof, rubber spatula or flat-bottomed wood spoon until it becomes as thick as pudding. Remove from the heat and pour into a 5 gallon (18.75 liter) bucket. Cover the bucket with a lid and store the paste in a cool place, for later use.

15.3: *A mortar mixer makes short work of fat plaster.*

The fat plaster recipe is intended as an example only. This recipe worked well for us on a particular project, but other methods or mixes may work better for you. After experimenting with the ratios for fat plaster, it's time to do some testing.

Making Plaster Test Patches

Keep track of the ratios of your various mixes. Slap the plaster onto the wall or a sample board. After the test patches have dried, examine the samples for cracking and strength. We dig into them with a screwdriver and bang on them with a block of wood. The same principles that apply to exterior plasters are also relevant to interior plasters (Fig. 15.4).

15.4: *Plaster test patches on the wall at Sand Island Ranger Station.*

If cracking is observed:
- Reduce the amount of water, *OR*
- Increase the amount of sand, *OR*
- Increase the amount of fiber, *OR*
- Try all three.

If the plaster appears chalky or weak:
- increase the amount of flour paste, *OR*
- use better quality clay.

We have found that with certain soils, the combination of the variety of fibers (particularly cellulose fiber) along with the flour paste, produces an unusual marbling effect that is quite beautiful. We suspect it is due to a chemical reaction among all the ingredients and the different drying rates of the fibers in such a thick layer of plaster.

Application Directly Over Earthbags: You Determine How Thick

Fat plaster goes on in two stages, the second one following the first immediately after the first stage has set up some, but is still green and a little pliable to the touch. (Please refer to the section entitled "Procedure for Hand Application Directly over Earthbags" in Chapter 14 for a detailed description of this process).

In hot, dry weather, work small areas at a time, like five-by-five-feet (1.5 by 1.5 m), so you can apply the second stage coat while the first one is still moist. Adjust your coverage to the conditions of the environment. After the plaster has cured to a leather-hard surface (still green, but firm), it can be hard-troweled and left as is, or rubbed down with rubber gloves or a sponge to raise the pores of the surface for a clay alis or milk paint application.

Mechanical Bond for Finish Plaster Coats

To prepare the surface for a subsequent fine finish plaster, the surface should be raked or otherwise roughed up by poking it with a nail board, scratching it with a coarse broom, or dragging a rake horizontally over it to add a toothy texture as a good key-in for the next coat. The surface is easiest to texture while still moist and pliable. (More information on mechanical bond can be found in Chapter 14).

Low-Fat and Fat-Free Plaster Over Chicken Wire

Low-fat plaster is a term we coined to describe a plaster with less fiber than fat plaster. The length of the straw and grass clippings need to be short enough to slip easily through chicken wire. Another option for low-fat plaster is to omit the straw or grass clippings and instead increase the paper cellulose and sand ratio. The idea is to have fiber that is fine enough to pass through the one-inch (1.25 cm) squares of the chicken wire, and fill the gaps behind the wire completely.

It follows then that a fat-free plaster is one that has no fiber whatsoever. Fat-free plaster is a good choice when the work is being done in a warm, humid climate where the drying process is very slow. Fiber-free plaster helps to decrease the occurrence of mold developing on the walls. Mold can stain and discolor the wall and create a possible health hazard.

Needless to say, this fat-free plaster will slip smoothly through and around the chicken wire squares. The wire needs to be strung up tightly against the bags without any saggy spots.

It can be applied by hand or with a trowel. Allow the plaster to set up some and then trowel on another coat about one-quarter to three-eights of an inch (0.625-0.9 cm) beyond the wire (Fig. 15.5).

15.5: *Fiber-free plaster goes on quickly, and deeply fills all the voids between the rows of bags until the plaster comes flush with the wire.*

Fat-Free Plaster Over Fat Plaster

We have had some mold appear on the surface of a few areas of walls using fat plaster. After completely drying, it easily brushed off, never to appear again. It was probably due to a lack of ventilation during the curing process. Since the presence of the organic fiber makes the plaster most susceptible to mold, the fiber can be omitted from the second coat application and used only in the first lath chinking coat. For walls that you intend to cover with milk paint, the finish will be smoother if the final surface is fiber-free. For the second stage cover coat, consider using a low-fat or fat-free plaster.

Omit any fiber and increase the sand ratio until the plaster dries without cracking. Experiment with the amount of flour paste, too. A fiber-free sand-rich earthen plaster is less apt to have mold problems. The mix works best when it is firm yet malleable and a little sticky. If there is adequate sand mixed with a stable, low-shrinking clay, even a thick fiber-free plaster will dry without cracking. Well-graded coarse, sharp sand is a prerequisite for success.

Rajuelas or Chinking Stones

The Anasazi Indians threw everything into their plasters: broken potshards, charred wood, bits of bone, grain chaff, cornhusks and broken cobs, yucca, and grass fibers. Mostly, though, they used small irregular stones pressed into the mud mortar in between the rock masonry walls. The Spanish called these little stones *Rajuelas*. Rajuelas reduce the shrinkage of the mud. In any case, filling the deep voids with a coarser gravel mix will reduce the risk of cracking, if achieving a single application, fat-free plaster coat is desired.

Types of Sand for Finish Earth Plasters

Fine finish plasters, in general, are just clay and sand that have been sieved through a finer mesh screen (Fig. 15.6).

The size of the sand dictates the thickness of the plaster. Fibers need to be both short and fine, reduced, or omitted altogether depending on personal preference. Casein binder or cooked flour paste can be added to keep the walls from dusting and give additional strength to a low clay plaster.

15.6: *We practice a fairly rustic style of plaster, using mostly a 1/8-inch to 1/16-inch wire mesh kitchen colander for screening our finish plasters.*

Ratios for clay and sand are typically 25-30 percent clay and 70-75 percent well-graded sand. Fine-screened wild-harvested clayey and sandy soils are our first choice for plasters, often amended with clean well-graded, fine sand. The cheapest source of fine sand is washed mortar-sand from a developed gravel yard. You can screen it further for a finer finish or use it as is. Super-fine quartz sand can be ordered at most lumberyards or pottery supply outlets, to make a very smooth refined plaster. One advantage of using bagged quartz sand is that the brilliance of the quartz sand enhances the color of a wild harvested clay, while a mortar sand will tend to darken or even slightly change the color of the naturally occurring clay.

Casein Fortified Finishes

A fat plaster made with an attractive clay soil is a beautiful finish on its own. To add further protection, we may want to brush on a coat or two of a clear *casein* milk glue. Casein is derived from milk protein. It forms the basis of a natural non-toxic binder that can be used as glue, a clear sealer, milk paint, and as a binder in fine-finish earthen plasters. Have you ever noticed the symbol of the Borden's milk cow on the label of Elmer's white glue? It's the original commercial milk glue. A casein binder is mildly water repellant, but still allows moisture to transpire through the walls, making them compatible with earthen surfaces.

We make our own casein binders from both fresh milk and processed, dried casein-powder (see the Resource Guide for sources of powdered casein). The following recipes produce the same result (for small batches, it's cheaper to use fresh milk).

Kaki's Homemade Milk Binder

This recipe makes one-half gallon (1.9 liters) of concentrated casein glue/binder.

Ingredients
- 1 gallon (3.75 liters) skim milk
- ½ cup white vinegar
- ⅓ cup borax (20 Mule Team Laundry Booster)

Heat the milk to about 110°-115° F (43.3°- 46.1° C); a warm hot. Take care not to overheat it or you'll cook the casein. Add ¼-cup of vinegar. Stir gently. The milk will begin to separate. Add the remaining ¼-cup of vinegar. Stir gently. The milk protein (casein) will glob up and the whey should turn into a clear yellow liquid (if it doesn't, just wait a little longer). When thoroughly separated, strain the contents through a wire mesh colander or cheesecloth. Gently rinse the globs of casein in cool water to remove the whey solution. Set aside.

Dissolve ⅓ cup of borax into 1 quart (or liter) of very hot water until the borax crystals completely dissolve and the water clears. Mix the casein glob with the borax solution and whip into a creamy froth, using a blender or manual or electric egg beater. Strain this casein/borax mix through a wire strainer into a bucket. Add one quart (or liter) of water to bring the whole mix up to one-half gallon (or about 2 liters) and stir thoroughly. You now have about one-half gallon of full-strength casein binder.

Variations

Casein binder can also be made from fresh milk using less vinegar than the above recipe, but it will take longer. Set a gallon of skim milk in a warm place in a bowl. Add a tablespoon of lemon juice, vinegar, soured milk, or yogurt. Allow this to curdle — it can take a day or two to separate. Strain it through a colander or cheesecloth. Since this is how panir cheese is made in India, at this point you can either eat it or follow the previous recipe to make casein binder.

Casein Binder Made from Powdered Casein

Mix together thoroughly: $1\frac{1}{3}$ cup of dry casein powder with $\frac{1}{3}$ cup dry, screened Borax. Whisk ingredients into half a gallon (1.9 liters) of water (you don't need a blender for this!). Allow 1 to 2 hours for this mix to thicken. That's it!

This half-gallon of casein binder can be used full strength as a base for making milk paint, or diluted further to use as a clear sealer or as a binder for a fine finish plaster over walls or an earthen floor.

Clear Casein Sealer

Casein can be used to provide additional moisture protection over an earthen plaster. Dilute half a gallon of casein binder with another half-gallon of water, bringing the total volume up to one gallon. The binder needs to be diluted to a consistency that will be thin enough, when brushed onto the wall, to soak in and dry without leaving a visibly shiny surface. Do small test patches on your wall or, preferably, a sample board, to determine whether more water will need to be added. In some cases, a full gallon (or more) of water may need to be added to achieve the desired results.

Clear Stains

To this casein sealer, earthen and mineral oxide pigments can be added to tint the binder to make a clear stain. Red iron oxide, yellow ochre, burnt umber, and ultra marine blue are some examples. Avoid acid dyes as they will curdle milk paint.

Milk Paint

This recipe will make one gallon of a basic-white opaque paint. Have all ingredients at room temperature.

- ½ gallon (1.9 liters) of full strength casein binder
- 8-10 cups white kaolin clay
- 8-10 cups whiting (calcium carbonate, limestone powder)
- 1½ cups titanium dioxide (white pigment)

Alternate, adding one cup at a time of the kaolin and the whiting, while whisking thoroughly until the consistency of rich cream. Mix the titanium dioxide with enough water to make a creamy toothpaste consistency. Whisk this titanium dioxide mixture into the creamy mix of binder, kaolin, and whiting. If the milk

15.7: Milk paint will dry eight to ten shades lighter than it appears when wet. Milk paint can be applied with a brush or roller, just like commercial paints.

paint thins down, add more kaolin. Do some paint tests to dial in the correct consistency. It should be thick enough that it doesn't drip off the brush, but thin enough to spread easily.

This recipe will make a brilliant, white, opaque paint. To make a warm eggshell cream color, add one level tablespoon of yellow ochre to the titanium dioxide powder. Adding one-quarter cup of the yellow ochre will make a color like fresh churned butter. Kaolin, whiting, and titanium dioxide can be found at pottery supply stores. They are usually available in 25- and 50-pound (11.1 and 22.2 kg) bags, or in smaller quantities.

Milk Paint Application

Wet milk paint appears transparent when first applied to a surface, yet will dry to a solid opaque. Sometimes, even over dark brown earth, one coat is enough, but usually two coats are better. Always apply milk paint at room temperature, as cold temperatures will thin it considerably. Refrigerate any unused portion. The paint will keep for about two weeks in a good, cold fridge. When ready to use, allow it to come up to room temperature. Coverage varies depending on wall surface but, on average, one gallon (3.75 liters) will cover about 150-200 square feet (14.5-19.3 sq. meters) on a smooth porous surface (Fig. 15.7)

Casein Binder Stabilized Finish Plaster

We live in the Painted Desert surrounded by a rainbow pallet of wild colorful earth. So, naturally we take advantage of what nature has to offer and collect our own plaster soils. These colorful clay soils are screened through one-sixteenth to one-eighth- inch (0.15-0.3 cm) kitchen colander screens into buckets. We then mix the screened clays with screened, washed mortar-sand (or quartz sand) at a 1:3 clay to sand ratio.

Dilute the concentrated casein binder to bring it up to a full gallon. This is the strength we usually use to mix with the clay and sand. The consistency is firm enough to sit on a trowel, yet still easy to spread. This plaster mix makes a fine finish coat on floors, too. Casein-stabilized plasters are very sticky — almost

gluey. What we like about them is that they add strength and a degree of water resistance without altering the subtle colors of the wild clays. Flour paste tends to darken the plaster and may even alter the color, which may or may not be desirable.

Water-Resistant Earthen Countertops and Bathing Areas

Gernot Minke's *Earth Construction Handbook* has examples of bathroom sinks and bathing areas sculpted out of "loam stabilized with casein" and sealed with linseed oil. As of this writing, these earthen bath fixtures have been in use for eight years. Casein-stabilized earth plaster can also be spread to make water resistant countertops and interior windowsills, and sealed with multiple coats of hot linseed oil and a few coats of a natural oil base floor sealer, as described in Chapter 16.

Alis

We learned from Carole Crews how to make a beautiful alis stabilized with cooked flour paste and store-bought kaolin clay and pigments. The same recipe can substitute finely screened wild harvested clay as well, but the flour paste tends to darken the original color. For those with limited access to wild clays or color selection, we offer a basic alis recipe, and contact sources for more in-depth finish techniques, from Carole and other fine *enharradoras*.

Recipe for Flour Paste Alis

Fill a 5-gallon (18.75 liter) bucket with 3 gallons (11.25 liters) of water and 2 quarts (1.9 liters) of cooked flour paste. Add equal parts of white kaolin powdered-clay and fine 70 grit silica sand or other washed, fine sand (Carole likes to add a portion of fine mica with the sand for a sparkly finish). Whisk the mixture into a smooth, creamy consistency. To help keep the clay in suspension, add a tablespoon of sodium silicate (a clear syrup that potters use to keep their clay slips evenly suspended). Add any premixed color pigments. Do test samples. The consistency should be thick enough to spread with a paintbrush or roller, without dripping.

The wall surface should have a suede-like porous surface to provide a good key in for the alis. It should also be dry and dust free. As with all paints and plasters, applying them from the top down keeps you from dripping on your finished wall surface. If two coats are necessary, allow the first one to completely dry before applying the second.

When the final coat is still leather-hard, it can be hard-troweled with a flexible stainless-steel trowel or, using small circular motions, a plastic lid from a yogurt container, as Carole Crews does. It takes some practice to master the yogurt lid technique (we're still trying to master it!), but give it a try. It is a less expensive alternative to flexible stainless-steel trowels.

Another finishing technique is to "soft sponge" the surface with a big, moist cellulose sponge, using circular motions to take out any brush strokes. This is also best done while the last coat of alis is still slightly green but firm. Always pick a small unobtrusive area to test for readiness before going whole hog on the entire wall. A coat or two of clear casein sealer can also further protect an alis finish.

Flour-Paste Fortified Finish Plaster

A similar recipe to the above-mentioned alis can be used to make a fine plaster as well. Simply bump up the ratio of the sand to 2½ or 3 parts. Use a well-graded finely-screened washed mortar-sand. Or consider using a variety of different sized quartz sand, like 30, 50, and 70 grit sand. Mix the consistency to a trowelable stiffness. Pre-moisten the wall before application. Allow the moisture to soak in — apply the plaster by hand and then trowel, or toss it on the wall with the trowel. Keep both your application hand and the trowel wet and slippery to encourage the mud to stick to the wall rather than your hands and tools.

Sealers

Linseed oil and BioShield's natural resin floor finish are about the only heavy-duty sealers we use over a poured adobe countertop stabilized with casein, interior windowsills or floors, or as a washable protective coat over a wainscot, like we did for the public visitor room at the BLM ranger station at Sand Island. (Fig. 17 in Color Section). Follow the same directions outlined in Chapter 16.

Some folks seal whole walls with linseed oil to maintain the wet look of the earth when it was first applied. In any case, a single coat of piping hot, boiled linseed oil can be used as a sealer/stabilizer over a fat plaster to inhibit mold and add extra resilience. Do sample tests!

Gypsum

Gypsum can be applied directly over an earthen plaster. We rarely use gypsum, so whatever we say about it would be secondhand at best. Therefore, we recommend reading *The Natural Plaster Book*, by Cedar Rose Guelberth and Dan Chiras, or checking your local library or the Internet for books by gypsum craftspeople. Remember — gypsum, like cement, is time sensitive, so its workability is limited to its set time, unlike earth that can be worked at your leisure.

Lime Finishes

(See Chapter 14 for a more in-depth discussion on lime).

We mostly use lime for exterior plaster, but there is no reason not to use it inside as well. As lime is a natural deterrent to mold, we may decide to apply a thin coat of limewater or lime wash over the surface of a fat plaster before proceeding with a fine finish earthen plaster.

Take care to dilute the lime wash enough so that it soaks into the surface of the plaster and gets a good bond. Many thin coats are better than one or two thick ones; otherwise the wash may curl away from the wall (a condition known as "potato-chipping"). Mixing a percentage of casein binder into the lime wash will keep it from dusting off. A half-gallon of concentrated casein binder mixed with one-half gallon (1.9 liters) of water can be used to dilute the lime putty, instead of water.

Lime plaster can also be used inside over a coarse earthen substrate and hard-troweled to a satin finish. Pigments to add for color must be alkaline-stable, like those used for staining cement/concrete, such as metal oxides and mineral pigments.

Fresco

For indoor walls, lime plaster can be a canvas for fresco by applying pure pigments mixed with enough water to make a paint, and applying them to a still green, "fresh" lime plaster. The pigment is literally sucked into the plaster by the carbonization process, causing the pigment to permanently bond with the plaster. Michaelangelo, look out!

CHAPTER 16

Floors

Adobe floors are a natural extension of earthen walls. Poured adobe or rammed earth floors offer a warm ambience and a resilient surface that can be polished to a glossy luster. The ease of work and finished beauty make them a natural for the first-time builder.

When we speak of floors in building terms, we usually think of a poured concrete pad, wood planks, or sheathing over wooden joists. However, once you have worked with a natural earth floor, you'll never go back. An earthbag building can accommodate any type of floor system, but our focus will be on earthen floors.

Earthen floors, whether poured adobe, rammed earth, or stone and mud mortar, are all built up from layers beginning with a capillary break, followed by an insulating layer, and ending up with a finish surface (Fig. 16.1).

There are infinite ways of finishing a poured adobe floor. What we present in the following pages is merely one variation. Let's start from the bottom and work our way up through each layer (Fig. 16.2).

16.1: *The beautifully finished patina of this earthen floor enhances the straw bale home of Kalen Jones and Susie Harrington in Moab, Utah.*

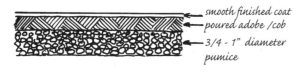

smooth finished coat
poured adobe /cob
3/4 - 1" diameter pumice

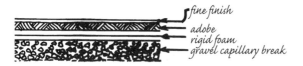

super fine color coat
semi smooth adobe
straw rich adobe
gravel capillary break

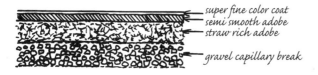

fine finish
adobe
rigid foam
gravel capillary break

16.2: *Three ways of insulating an adobe floor.*

16.3: *Rake this gravel out level and the lowest layer of your earthen floor is done.*

Sub-Floor: A Capillary Break

Just as the walls and roof protect the interior of a building and its inhabitants from the vagaries of the weather, so a floor protects the building and occupants from the whims of the exposed earth. One of the biggest ravagers of an earthen floor is moisture rising up into it. This movement of water upwards through a porous surface is called capillary action. To avoid it, a capillary break is installed between the earth and the finish floor. Gravel is the simplest and yet most effective capillary break.

Rake and tamp the excavated floor as level as possible. We use our hand tampers, but a large space can be tamped most quickly with an electric compactor, like the kind used for the bed of a sidewalk. Though it's not imperative to make it level, it is easier to make the succeeding layers level if we start from level.

We then spread washed, three-quarter-inch (1.875 cm) gravel at least four inches (10 cm) deep. Large-pore spaces, like the air spaces in gravel, prevent the water from migrating upwards. Instead, gravity takes over, keeping the water from climbing up. The gravel is our precaution against moisture wicking up from the ground (Fig. 16.3).

Mid-Layer: The Comfort Zone

For a non-insulated earthen floor, we could proceed to pour our adobe right on top of the gravel. Keep in mind that the earth below frost level maintains a fairly constant temperature of 52-58° F (11-14° C). This can make for a comfortable floor in the hotter summer months, but in the colder winter months, can result in a floor that is uncomfortably cold. For this reason, it's to our best advantage to insulate against this cold seeping through the floor.

For a low-tech insulated floor, we mix a high ratio of straw to clay adobe. In order to prevent the adobe from penetrating too deeply into our gravel base, we first spread a two-inch (5 cm) layer of loose, clean, dry straw (Fig. 16.4).

In general, a mix that contains 25 to 35 percent clay-rich soil, and 65 to 75 percent sandy/gravelly soil, like reject sand or road base, is combined with as much long straw as the mix will accommodate and still feel fairly stiff, but not dry. This is often an occasion for a mud-mixing party, with much stomping and thrashing about (Fig. 16.5).

This straw-rich base layer goes on about four to six inches (10-15 cm) thick, depending on your stamina.

Screed or trowel this layer as level as possible. One way to help maintain a level surface is to use stringlines or partition the floor into sections with boards that delineate the height of your pour (Fig. 16.6).

Leave the surface textured to provide a keyin for the next layer. Though it seems obvious, it bears reminding to work your way towards a door, rather than trapping yourself up against a wall, with no way out except across the floor you have worked so hard to level.

16.5: *Mixing a straw-rich adobe insulative layer can be likened to shampooing a big shaggy dog.*

16.4: *The loose straw soaks up any extra moisture from the adobe and provides that much more insulation.*

16.6: *When one section is poured and leveled, the boards are moved and the adobe poured into the new section. Continue this method until the pour is complete.*

16.7: *Finish floor troweling tip: A padded kneeling board protects both the floor and your knees. Supporting your weight on a wood-float frees your other hand to trowel with just the right amount of pressure.*

This high-straw pour will take days to dry in hot, dry weather and longer in a more humid climate. It could take weeks in rainy or cold weather. In this case, to inhibit mold add a cup of Borax to every wheelbarrow load (see more on this natural mold inhibitor in Chapter 15, under "Additives"). We use a cement mixer for all our earthen plasters, cob mixes, and floor pours. Cement mixers are cheaper and easier to come by than mortar mixers. A cement mixer can also handle coarser material, like gravel and long straw.

Once the straw/clay base is dry, it should sound hollow when you knock on it. If it doesn't, it may not be dry all the way through, or the ratio of mud to straw could have been a little off. A hollow sound indicates lots of air spaces, meaning an insulative success. It's imperative that this insulative layer is completely dry before continuing on to a finish. Complete drying allows time for any shrinkage that might occur. Cracks may occur during this drying process, but the next layer will fill those cracks in, and the small cracks provide a key-in for the final coat. If the insulative layer is not dry before you apply the final coat, cracks can occur that travel up into and through the finish coat. So let it dry completely to save you from extra work and unnecessary frustration.

Final Layer — The Beauty Coat

After thorough curing of the insulative layer, a finish layer can be poured. This layer is a fine mix of screened earth with about 25 percent clay-rich soil and 75 percent sandy soil passed through a one-quarter-inch (0.625 cm) screen. Chopped straw can be added to this mix if desired, but the straw should not exceed 1½ inches (3.75cm) in length. This mix feels best when it is moderately firm, yet still easy to spread. Adjust the mix by experimenting with the ratios until the test patch dries without cracking. Spread this coat about three-quarters to one-inch (1.875-2.5 cm) thick. Before applying each layer of adobe, pre-moisten the sub-layer with water. Allow it to soak in, and then you're ready to spread the next layer. The moisture helps reactivate the clay to provide a good bond between the two layers.

Any fine cracks that do develop can be filled with a finer mix of screened earth up to one-sixteenth to one-eighth inch (0.15-0.3 cm) thick. The size of the sand dictates the thickness of your final finish. You can go on forever applying even finer layers (Fig. 16.7).

You can omit the straw in this final coat or add a few handfuls of fine-chopped quarter-inch (0.625 cm) long straw. This is all personal preference; as to how *you* want your finish coat to appear. Pigment can be added to this layer to create a distinctive color or pattern.

Mud-Mortared Stone or Saltillo Tile
(Fig. 16.8)

If you want to have stone or tile as part of a finish level, proceed as explained for the initial two layers: the gravel followed by the insulative layer. When that is completely dry, mix up a batch similar to the described final layer. Rather than spreading and leveling it, plop it into place where you want the stone or tile to go. Work the stone or tile into the mud mix so that it remains at least a quarter-inch (0.625 cm) above the level of the mud mortar. Level the stone and then plop on more mortar and your next stone. Level this stone or tile with the last one, and continue as above. Sometimes the use of a rubber mallet helps persuade a stone to settle into the mud mortar better.

As you get further from your original stone or tile, it becomes necessary to level the succeeding ones in all directions to obtain a level surface. Once all of your stone and/or tile is set in place and level, allow it to dry completely before continuing to the next phase: the grout.

Grout

Earthen grout is like-commercial grout used for conventional tile work. It fills in the spaces between the tiles or stones creating a level surface. The advantage of earthen grout is that it is much less expensive and contains none of the harsh chemical additives found in commercially manufactured grouts.

Begin with the mix outlined above for a finish layer. Use a mix that contains about 25 percent clay-rich soil and 75 percent fine-screened sandy soil. The smaller the screen the better, and try not to exceed one-eighth-inch (0.3 cm). No straw is added to this mix. It should be softer than that mixed for the mortar to set the stone. It should be smooth enough to work into the joints between the tile and stone, but not so loose as to be runny. Using a sponge float or damp cellulose sponge, work the filled joints with a circular motion. This helps force the grout into the joints and drives out any air pockets. Add more grout to an area that is lower than the set stone/tile. We are trying to achieve a level surface, so a lot of very fine material will be floated onto the surface of the stones. This is not a problem as the stone/tile can be easily cleaned with a damp rag later. To create a water resistant grout, instead of adding water to your dry grout mix, try using a casein binder as your wetting agent.

Casein-Stabilized Earthen Grout

We used a red clay/washed sand casein-binder grout mix to fill the voids between the randomly set tiles and stone in the floor of the Honey House dome. The morning we entered the dome to seal the floor with hot linseed oil, we were shocked to discover seven inches (17.5 cm) of red, muddy water floating in it. There had been a flash flood the night before that had traveled across our property and poured into the dome.

16.8: *Mud-mortared flagstone and antique Malibu tiles grace the floor of the Honey House dome.*

What impressed us was how well the casein-stabilized grout resisted water penetration, even after sitting overnight. We bailed and sponged out the water and found the grout was a little soft for the first eighth of an inch (0.3 cm) or so, but underneath was solid. If not for the casein, the water would have soaked right through into our straw/clay insulated sub-floor.

A full recipe for making casein binder for a variety of applications can be found in Chapter 15 under the sub-section entitled "Additives".

Sealing An Earthen Floor

Since floors receive a lot of heavy traffic, an earthen floor will wear down over time and make a lot of dust, unless it is sealed. Our choice is the non-toxic natural route that will enhance the beauty of the earth while providing a level of hardened protection. The most commonly used sealer on an earthen floor is linseed oil. Properly applied, linseed oil can add not only a degree of hardness, but also provide water resistance, allowing an earthen floor to be washed. The trick in putting linseed oil on an earthen surface is in the application.

The more deeply linseed oil penetrates the earthen surface, the better protection it provides. There are two ways to accomplish deep penetration. One way is to cut the linseed oil with a proportion of thinner. We prefer to use a citrus rather than a petroleum-based thinner due to the hazards associated with petroleum products, particularly in an enclosed environment. Even with a citrus thinner, allow for plenty of ventilation while it dries.

Linseed oil will considerably darken the color of the dry earth it is applied to. It will also change the color somewhat — reds will turn dark brown, etc. To try and maintain as much of the original color as possible we mix our final troweled-on finish plaster coat with casein binder. The casein binder seems to add a level of color protection while still allowing the oil sealer to penetrate. To be certain, try out a test patch in a discrete location to see what works best.

Typically, an earthen floor receives at least three coats of oil. This allows for deep penetration into the earthen surface. There are two schools of thought on how the oil and thinner should be mixed and applied. One way is to go from thinner to thicker. That is to say, the first coat should be about 75 percent citrus thinner and 25 percent linseed oil. The second coat goes on at 50:50 and the third coat is 25 percent thinner to 75 percent oil. Any further coats are cut even more. Let each coat dry completely before applying the next. This takes anywhere from 24 to 48 hours, depending on the relative humidity.

Frank Meyer creates tamped earth floors in Texas. He claims that because of the humidity, the oil coat should go on the opposite of what we described above. The first coat is cut 25 percent thinner to 75 percent oil, and each successive coat receives more citrus thinner until the final coat is at least 75 percent thinner to 25 percent oil. Bill and Athena Steen create beautiful straw bale structures with earthen floors. They too suggest starting thick and finishing thin, even though they live in the arid Southwest. An alternative is to use a third method that works well in both arid and humid environments.

Rather than adding a percentage of thinner, we heat the oil in a double boiler. This naturally thins the oils, which allows it to penetrate into the earthen floor. What we have found is that even by the third or fourth application, the warmed oil is still penetrating. Often with citrus thinned oil, the third coat begins to set on the surface in places without penetrating further, leaving a tacky residue. This doesn't happen as quickly with the heated oil. As the citrus thinner is quite strong, this warm oil application is friendlier to work with and seems to dry quicker with less smell.

No matter which way the oil is put onto the floor, it is applied in the same manner. We use a large, soft paintbrush to apply the oil, spreading it evenly and working it into the earthen surface. After two or three coats, the earthen floor takes on a degree of hardness and water resistance, but just a degree. We could stop here and simply perform periodic maintenance on the floor by applying extra coats each year or as needed. But, as mentioned earlier, sometimes some of the oil sits on top and remains tacky. To alleviate this and create a hard, cleanable, and (if desired) waxable surface, we like to add one more step.

This final step involves applying a natural sealer over the oiled floor. We use a natural oil-based floor finish that is designed for wood, but works equally well on earthen surfaces. This floor finish is called Natural Resin Floor Finish and is made by Bio-Shield Paints (see the Resource Guide under "Bulk Oxide Pigments"). Two to three coats of this finish set up hard enough to be washed and even waxed. Two coats have lasted on our Honey House earth and stone floor for years without any further maintenance.

As with applying oil or a thinner/oil combo, allow each coat to dry completely before continuing with any successive coats. This can take as little as 24 hours per coat if the weather is warm and dry, or as much as 48 hours between coats in damp or cold weather. It is better to wait a little longer to make sure the preceding coat is completely dry before continuing. For best results this hard finish coat should be applied by brush, working the sealer in all directions and applying it thinly. This sealer works equally well on earth, stone, and wood. Although pricey, its compatibility, ease of working with, and beautiful final result are well worth the expense.

A properly constructed and finished earthen floor is unequaled in beauty and comfort. Start small and work your way up to larger projects. When someone comments on the beauty and practicality of your earthen floor, the time that went into preparing it seems insignificant. The materials for making an earthen floor are often site-available and easy to acquire, with a bit of effort on your part. Your endeavor will be rewarded with the knowledge that you created something of beauty with your own hands that is comfortable and kind to our environment.

CHAPTER 17

Designing for Your Climate

Natural interior climate control of a building is an art and a science that takes in a variety of design considerations. What we offer is basic common sense examples that are simple, affordable, and low-tech. We suggest taking a class in permaculture for site and passive solar orientation for your locale.

The book *Alternative Construction*, by Lynne Elizabeth and Cassandra Adams, offers case scenarios for six different climate zones and which hybrid combinations of mass and insulation suit each climate. An expert can customize the appropriate proportion of insulation to mass, but by looking at nature and the

17.1: *Shade is a priority in a hot, sunny locale, and extends the living space outdoors.*

dwelling devised by indigenous people, we can get a feel for what design features we too would find comfortable.

Let's look at design strategies for cooling, warming, and keeping an earthbag home, dry, that can be adapted to suit a range of climates.

Strategies for Keeping Cool

Wall Mass

In the desert, where cooling is a priority, thick earthen walls do an excellent job of maintaining a pleasant

17.2: *Lime whitewashed walls reflect the sun's intensity in sunny climates.*

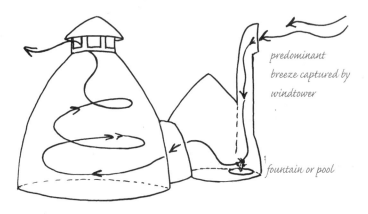

predominant breeze captured by windtower

fountain or pool

17.3: *Wind-catchers.*

interior temperature. The key word here is *thick*. Exterior temperatures penetrate an earthen wall to a depth of 12 inches (30 cm) before the momentum of the external temperature is dissipated. When simply relying on mass alone to moderate internal temperatures, consider using the wider 18-20-inch (45-50 cm) 100-lb. bags, or two rows of 12-inch (30 cm) tubes side-by-side, etc.

Daily Flywheel Regulation

Summertime in the high desert, where days are hot and nights are cool, we open all the windows and doors before we go to bed to invite the cool air inside. In the morning, we close the building back up. The cool air captured by the internal mass is slowly released into the living space during the day. By evening, the warmth of the day catches up. Soon, the outside air has started to cool off again and we let it back in. This is a simplified example of a 24-hour flywheel effect.

Shade

Long overhangs shield the walls from direct sun. Wrap-around porches, trellises, vines, and trees help to keep the walls from heating up (Fig. 17.1).

Exterior Color

The exterior color of the walls (or the roof of a dome) significantly affects the surface temperature of a building. The darker the color, the more heat is absorbed, whereas a lighter color reflects the heat. A white-washed surface reflects upwards of 70 percent of the sun's radiation. This is why people engage in the yearly application of lime wash in many sun-baked Mediterranean countries, like southern Italy and Greece (Fig. 17.2).

Ventilation

In the Middle East, tall chimney-like structures called *wind catchers* or *wind scoops* are used to catch the predominant breeze and funnel it down into the living space, often passing it through a small fountain or pool to pre-moisten the air (Fig. 17.3).

STRATEGIES FOR KEEPING COOL

Some strategies for keeping cool include:

- Sufficient wall thickness
- Wall shading: deep overhangs and porches
- Greenery: trees, trellises, vines
- Living or thickly insulated roof
- Ventilation: wind catchers
- Light colors for exterior wall and roof surfaces
- Regulation of daily flywheel
- Berming, burying, or digging in
- Exterior insulation
- Window shades

Wind catchers are the original passive cooling systems that have been at work for centuries — before the introduction of mechanized evaporative coolers and air conditioning.

Living Roofs

If enough water is available, living roofs are one of the best roof coverings for keeping the most exposed surface of a building cool. Tall grasses, succulents, or cactus plants provide shade and moisture over one's head.

STRATEGIES FOR KEEPING DRY

- Big overhangs
- Tall stem walls

"A good hat and tall boots keep an earthbag home dry and healthy."
— Mr. Natural

17.4: Long overhangs, porch roofs and tall stem walls keep an earthbag home dry and healthy

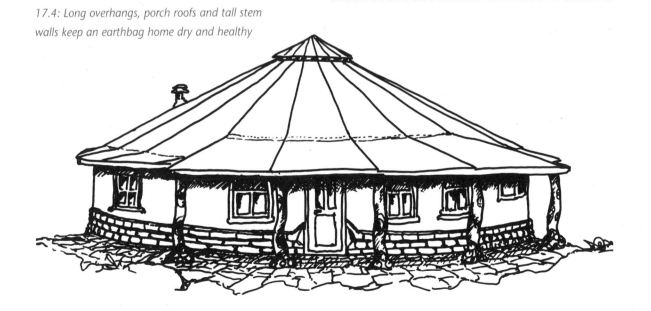

Passive Solar Gain

Traditional earthen architecture limits the invasion of direct sun through large glass openings for a couple of reasons. First, earthen buildings rely on sufficient mass to provide stability. Big windows mean less wall mass. Second, glass is a relatively recent invention compared to the thousands of years dirt architecture has been around.

Earthen walls act as a buffer from the assault of the sun in summer, and as an external heat sponge for absorbing the low-angled sun in winter (Fig. 17.5).

To get the benefit of modern day passive solar design, while making the best use of an earthen structure's mass, add a wraparound sunroom, or a greenhouse built of wood framing that can accommodate a lot of glass — rather than to risk compromising an earthen wall with a series of big windows. Sun entering the glass room will heat up the earthen walls, which act as a thermal storage bank. Later in the day, as the sun retires, the heat is slowly re-radiated back into the living space. Cold, sunny climates are prime areas for this strategy. Cold, cloudy climates will need to supplement their heating with auxiliary systems, i.e., wood burning stoves, radiant floor heating, gas furnace, etc. (Fig. 17.6a & b).

17.5: *Angle of exposure for summer sun and winter sun.*

An enclosed sunroom creates a buffer from the external environment. This is easy to regulate by closing it off from the rest of the house or venting it in warmer weather. Insulated window shades keep heat from escaping at night. It is much easier to regulate the heat generated within an attached sunroom than within a whole house built with a lot of south facing glass.

Insulation Strategies for Earthbag Walls

Exterior insulation helps to make more efficient use of earthbag mass by creating an air buffer of resistance to extreme external temperature changes. Here we offer a few techniques for attaching various insulative materials to earthbag walls, from minimal R-values for moderately cold climates to mega-insulation for long, bitterly cold winters.

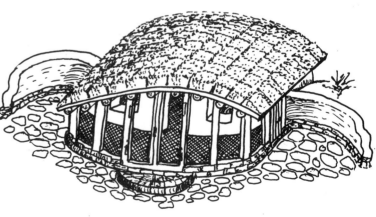

17.6a & b: *Bermed oval-shaped vaulted viga earthbag cottage with enclosed wraparound sunroom and living straw bale roof.*

Rigid Foam

Rigid foam can be screwed into a quality rammed-earth soil. What bugs us about rigid foam is that, first, it's foam, and second, it doesn't breathe. For below ground applications we don't expect the walls to breathe, but for above ground walls, it seems silly to build these lovely, natural earthen walls and then suffocate them in plastic. If rigid foam seems essential, consider drilling the foam full of one-quarter-inch (0.6 cm) holes, spaced six inches apart in every direction. Then attach it to the earthbag structure, using long screws that penetrate at least two inches (5 cm) into the earthen fill (Fig. 17.7).

Chicken wire will also need to go over the foam to provide a key-in for plaster. Screwing into the walls risks fracturing the compacted earth, and so much screwing around compromises FQSS principles.

Perhaps there is an alternative, natural, breathable straw board version of rigid foam we just don't know about — one that offers adequate insulation and is flexible enough to contour to earthbags. Any ideas?

Spray-on Paper Adobe

An alternative approach is a spray-on application of paper adobe, or lime-stabilized paper adobe, directly over the exterior walls. Waste paper in the US takes up the largest volume of space in our landfills. Our local Utah State Job Service office generates several huge garbage bags full of shredded paper each day! Here's a tip for scavengers without recycling programs in their area: throw all those damn slick color catalogues into the mixer with some clay and sand and spray them on the walls nice and thick (2-3 inches [5-7.5 cm]). Seal it with a lime plaster.

Some pioneers claim to be getting an R-value of 3.5 per inch (2.5 cm) of papercrete block. There are several pioneers experimenting in the field of papercrete and paper adobe slurries producing strong, lightweight, insulative blocks, panels, and plasters. (Please consult the Resource Guide for more info).

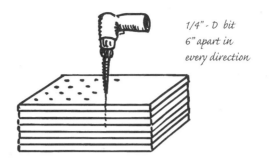

1/4" - D bit
6" apart in
every direction

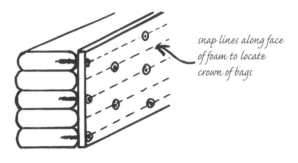

snap lines along face
of foam to locate
crown of bags

fasten foam with galvanized deck screws 2" - 2.5" deep

Use plastic roofing-washers or wind locks over screws
for the most secure connection

17.7: Prepping and attaching rigid foam insulation onto an earthbag wall.

Pumice/Scoria Earthbag Walls

If you live where either pumice or scoria is available, building the walls with up to 50 percent of either of these, with the balance of the mix compactable dirt, may add a degree of insulation. We've made both rammed-earth pumice bags and slurried adobe/pumice bags at a 50:50 ratio of binder to pumice (see "Insulated Earthbag Foundations/Stem Walls" in Chapter 4). These bags could just as easily be used to build whole walls, as they are still solid and strong, but weigh only 60 pounds (27.2 kg) as compared to the usual 90-100 pound bag (41-45 kg).

Mega-Mass Meets Mega-Insulation: Earthbag/Straw bale Hybrid Wall

Straw bales offer excellent R-values: 35-45 depending on size, compaction, and who you are asking. They are also easy to attach to an earthbag wall. The following are some illustrated examples of configurations of straw bales married to earthbags (Fig. 17.9).

Considering two- to three-foot (0.6-0.9 m) thick walls are common for traditional earthen structures throughout the world, adding straw bales to the outside of an earthbag wall would seem sensible. We can take advantage of the benefits of the earth's mass (U-value) and the straw bale's insulation (R-value) to build a home that will be effortless to heat and cool, using two low-tech systems together (Fig. 17.10).

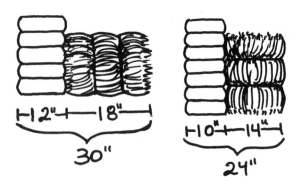

17.9: *Alternate configurations of earthbags and straw bales used to achieve mass and insulation.*

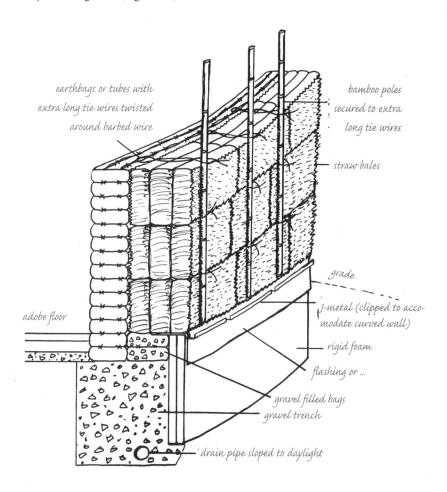

17.10: *Earthbag wall with straw bale wrap.*

Interior Earthbag Walls with Exterior Strawbale Walls

Another hybrid version is to split up the mass and the insulation by using the earthbag walls as interior partition walls and the strawbales as exterior insulated buffer walls (Fig. 17.11 a & b).

17.11a (right): 1 1/2 story earthbag interior structure enclosed with exterior insulated straw bale walls.

17.11b (below): This drawing depicts design strategies for using multiple materials; earthbags as interior walls for absorbing warmth from an attached sun room and straw bales as exterior insulating walls.

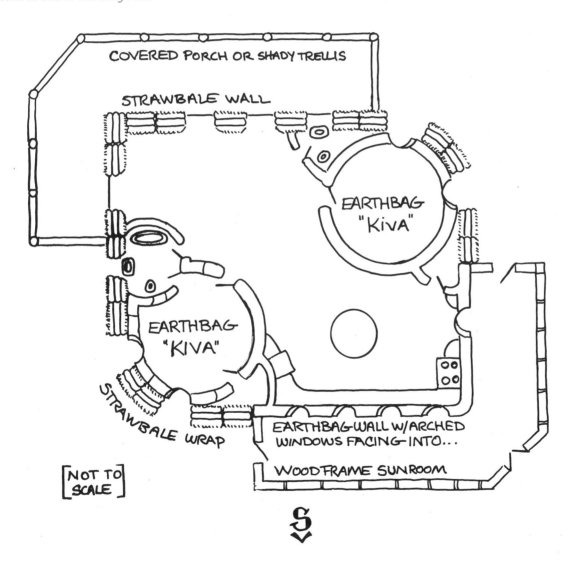

COVERED PORCH OR SHADY TRELLIS

STRAWBALE WALL

EARTHBAG "KIVA"

EARTHBAG "KIVA"

STRAWBALE WRAP

EARTHBAG WALL W/ARCHED WINDOWS FACING INTO...

WOODFRAME SUNROOM

[NOT TO SCALE]

Straw bale Walls with Earthbag Foundations

Earthbags can be used to build foundations for straw-bale walls. Extra care should be taken to prevent moisture wicking up into the bales (Fig. 17.12 & 17.13).

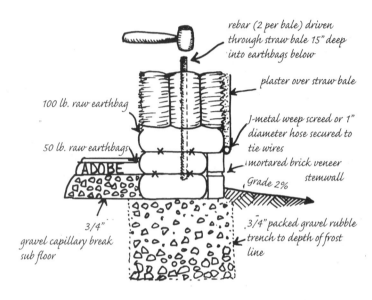

rebar (2 per bale) driven through straw bale 15" deep into earthbags below

plaster over straw bale

100 lb. raw earthbag

50 lb. raw earthbags

ADOBE

J-metal weep screed or 1" diameter hose secured to tie wires

mortared brick veneer stemwall

Grade 2%

3/4" gravel capillary break sub floor

3/4" packed gravel rubble trench to depth of frost line

17.12: Grade level earthbag foundation for non-load bearing straw bale walls.

17.13: Insulated below grade level earthbag foundation for straw bale walls.

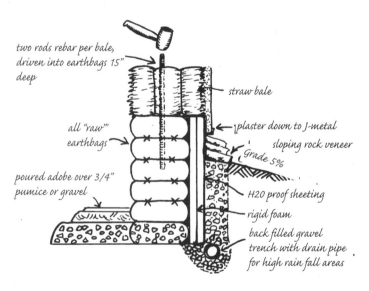

two rods rebar per bale, driven into earthbags 15" deep

straw bale

all "raw" earthbags

plaster down to J-metal

sloping rock veneer

Grade 5%

poured adobe over 3/4" pumice or gravel

H20 proof sheeting

rigid foam

back filled gravel trench with drain pipe for high rain fall areas

Advantages of Bermed and Buried Structures

One of the big advantages of using earthbags is that it is a simple, inexpensive, low impact wall system for building below the ground. We often hear rabid insulation enthusiasts declare earthen architecture inappropriate for cold climates due to its lack of insulative characteristics. This is true if you are building a house that sits on top of the ground fully exposed to the elements. But like many animals that hibernate, people in the coldest of climates did the same: snuggled into the earth (Fig. 17.14).

By burrowing into the earth we reduce the temperature extremes to a moderating 48°- 55° F (the average temperature of the earth below frost level). Using the earth's warmth means we can use a minimal amount of exterior buried insulation. Even a straw bale or wood frame structure can take advantage of a bermed north wall or a basement built using earthbags instead of concrete. The surrounding earth acts as a natural temperature regulator for both cold and hot climates. David Pearson in *The Natural House Book* explains: "The soil, depending on its depth and thermal properties, slows the passage of heat gained or lost to such an extent that the heat gained in the summer will reach the house in early winter, and the cooling effects on the soil in winter will not flow through to the house until early summer."

A bermed/buried structure means less exterior wall surface to finish, and provides easy access to the roof. The lower profile integrates nicely into the landscape, and any excavation work provides building material for some or all of the intended earthbag walls, plaster, or rockwork.

Well-ventilated earthbag domes excel for subterranean living due to their structural integrity, and for food storage due to their consistent temperature. When we step down into our Honey House

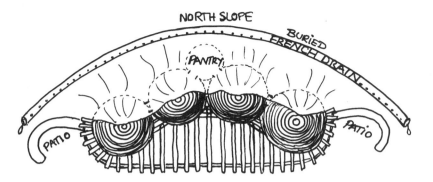

17.14: *Four interconnected, bermed domes with buried dome pantry flanked by retaining walls — south face enclosed with attached sunroom in cold climates, or use as covered porch or trellis in hot climates.*

dome, there is a hush that follows, with an air of solid reassurance. Of course, different people are attracted to different living environments. All we are saying is that earthen and, in particular, earthbag architecture can be adapted to a cold climate, given the necessary attention to detail and design (Fig. 17.15).

STRATEGIES FOR KEEPING WARM

- Ample southern exposure
- Ample exterior insulation
- Super-insulated roof
- Attached sunroom or greenhouse
- Dark floor surfaces in sunroom
- Insulated curtains
- Proper length eaves for the latitude
- Dark exterior wall surfaces
- Efficient auxiliary heating system
- Lots of close family and friends
- Big animals
- Bermed or buried floor level

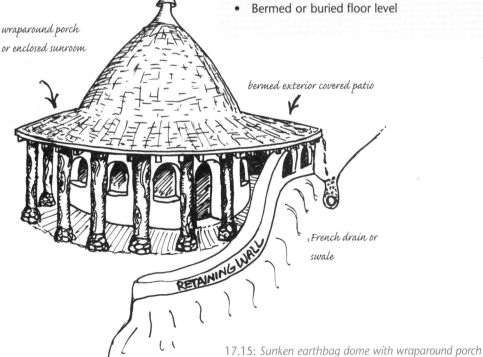

17.15: *Sunken earthbag dome with wraparound porch*

CHAPTER 18

The Code

John and Jane Doe worked hard for years to save enough money to buy a small piece of property where they always dreamed of living. Jane was chemically sensitive to many of the manufactured materials that go into the construction of new structures. For these reasons, purchasing a pre-manufactured home to put on their property was out of the question. Besides the toxicity of such structures, their cost was prohibitive, and to have a contractor build a wood frame house was beyond their financial means.

They both loved the outdoors, were unafraid of hard work and were rather handy, too. They investigated various forms of alternative architecture, attended a couple of natural building workshops, and helped friends with their small construction projects. They felt confident that if they could discover a building method that was simple yet durable, they had the self-assurance and just enough money to attempt a modest-sized structure on their own.

John held down two jobs, but still found time to volunteer with a local recycling project where he learned the value of reducing his needs for non-renewable commodities. Jane's chemical sensitivity led her to investigate the timber industry's practices. She was appalled to discover the amount of energy, waste, and unpronounceable chemical compounds that go into the production of wood-based materials. The deforestation of old-growth timber in her own country and the ever-increasing loss of equatorial and tropical rain forest land throughout the world saddened and dismayed her. For

these reasons, they decided to build their home with the least amount of wood possible. After investigating several types of alternative building methods, they settled on the idea of building with earth.

They carefully examined their property to determine the best location for the proposed structure, paying attention to site orientation in order to facilitate energy efficiency, good drainage, and access, with the least amount of disturbance to the natural lay of the land. On paper, they prepared a basic floor plan that outlined their family's needs and was adequately sized while remaining compact enough to be affordable to build, heat, and cool. When they were satisfied with their preplanning, they went to the county courthouse to apply for a permit from the building department.

When they arrived at the building inspector's office, he warmly greeted them and offered them seats. He listened attentively as they explained their situation and what they intended. When they described to him their desire to build with earth, he nodded agreeably and directed them to some shelves in his office lined with books containing information on various types of building methods and mediums.

Knowing that John and Jane were particularly interested in earthen construction, he directed them toward the part of his library concerned with earth building. He showed them different earth building techniques that were particularly suited to the climate and soil type most commonly encountered in their area. He was able to show them the advantages and *disadvantages of various kinds of earth construction, particularly in terms of what they could most reasonably afford and accomplish themselves.*

When they showed him their simple floor plan, he patiently explained to them some common pitfalls first-time builders may encounter and how to best avoid or alleviate them. Understanding their financial situation, he encouraged them to think about building something smaller and simpler, while planning for possibly adding on later as the need arose and finances allowed. His enthusiasm and willingness to share his accumulated knowledge left the Does feeling inspired and encouraged.

As he walked them to his office door he assured them that he would be available to them if they needed advice or questions answered. "After all," he amiably quipped, "I am a civil servant entrusted with the responsibility to protect your public health and safety."

The preceding story is interesting in that it is both factual and fantastical. All too many people are familiar with the beginning of the story. Thousands of people in this country, not to mention millions throughout the world, suffer from the buildup of toxins in the environment and the inability to afford their own home.

The exponential degradation of our environment and the increasing disparity of the haves and have-nots is a social dilemma precipitated by the preponderance of bureaucracies in the industrialized world. The building codes as they exist today are an extension of this bureaucracy; the perceived need for control of how and what we build takes the power from the many and transfers it to the few. This was not the original intent of building codes. For a better understanding of what the codes represent and how they have evolved, let's take a brief look at their history.

The earliest known building code comes from Babylon in the 18th century B.C. Its intention was to protect the household from death or injury brought about by shoddy workmanship. The code was designed to inflict upon the builder the same fate that the occupants suffered as a result of inadequate building procedures. An eye for an eye was the law

of the land. In ancient Rome, building engineers were required by law to stand beneath a completed arch as the formwork was removed. It is not surprising that so many Roman arches still stand today.

In 1189, London enacted a law that required common walls between separate structures to be of masonry construction. The purpose of this law was to prevent the rapid spread of fire from one building to another. The earliest enactment of codes in the US was also related to the incidence of catastrophic fire in urban areas. In 1630, Boston passed a law that prohibited the construction of wooden chimneys and thatched roofs.

As more immigrants poured into the US, overcrowding occurred in cities along the Eastern seaboard. This overcrowding, coupled with inadequate sanitation, posed a health threat to the occupants of these early tenements and the cities that hosted them. The Tenement House Act was introduced in New York City in 1867, to counteract the intolerable living conditions of these structures. Among other things, this ordinance called for one water closet (or privy) for every 20 occupants.

It is easy to see from these early examples that the scope and intent of building codes was to protect the occupants and general public from threats to life and limb resulting from shoddy workmanship, inadequate fire suppression, and pathogenic sewage conditions. So what happened along the way to turn these well-intentioned laws and acts into the oppressive, inflexible rules and regulations that restrain the rights and liberties of the very people they were originally designed to empower?

By the beginning of the 20th century, building regulations were applicable only in the larger cities of the US. In 1905 the first national building code was penned, prepared by the National Board of Fire Underwriters, a group representing the insurance industry. They proposed a nation-wide building ordinance that would minimize their risks and cut their financial losses. They were so successful that other self-interest groups saw the self-serving gain of legal control over building construction. In 1927, a group calling themselves the International Conference of Building Officials (ICBO) gathered in Phoenix, Arizona, to prepare and sponsor legislative enactment of the Uniform Building Code (UBC). Perhaps it is only coincidental that these self-proclaimed building officials were primarily comprised of building material suppliers and manufacturers, labor organizers, and other building professionals.

Professional societies, insurance underwriters, lending institutions, trade associations, labor unions, and contractor associations have all had a special interest in influencing the code. The proliferation of building regulations promoted by influential groups largely from the private sector continued willy-nilly until by 1968 there were nearly 5,000 different codes in the US alone. As of this writing, the four most commonly used codes are the National Building Code, in the eastern US, the Uniform Building Code in the western US, the Southern Standard Building Code in the southern US, and the Basic Building Code in the remaining states that even have a code. They all contain virtually the same information. Aside from these building codes, other duplicative codes exist concerning electricity, elevators, fire-prevention, plumbing, mechanical, housing and a host of other miscellaneous codes.

As early as 1921, it was recognized by many, including a Senate Committee on Reconstruction and Production, that building codes contributed to unnecessarily high construction costs. In 1970, the Secretary of the Department of Housing and Urban Development, George Romney, said that 80 percent of the American people could not afford to buy code-regulated bank loan-approved contract-built housing.

Among the plethora of guidelines, rules, laws, ordinances and amendments written, no specific standards exist for the use or application of alternative materials and methods. Wood framing, concrete, and steel are comprehensively covered in agonizing detail, yet there is no mention of centuries-old tried and true forms of building, like earthen arches, domes, vaults, or even the common post and beam construc-

tion. The only mention of alternative building methods and materials within the UBC is located in Section 104.2.8:

> . . . The building official may approve any . . . alternate, provided the building official finds that the proposed design is satisfactory and complies with the provisions of this code and that the material, method or work offered is, for the purpose intended, at least the equivalent of that prescribed in this code in suitability, strength, effectiveness, fire resistance, durability, safety and sanitation.

This clearly indicates that the building official has the power to approve any alternate system or material. Furthermore, Section 104.2.6 of the UBC exonerates the building official from any personal liability if he/she is "acting in good faith . . . in the discharge of the duties required by [the] Code."

When an alternative form of construction is proposed to a building official that he/she is unfamiliar with, "the building official may require tests as proof of compliance" with the code. There is a Catch-22 attached to this seemingly innocuous requirement; "All tests shall be made by an approved agency." These tests cost thousands of dollars that the tightly budgeted owner/builder can rarely afford. Approved test facilities are few and far between and constantly backlogged. Ironically, the code does not acknowledge the ultimate test of any building form, the test of time. A construction method that has existed for centuries, with structures still in use after hundreds of years, is of no consideration to the code. To further complicate this issue of testing, one jurisdiction does not have to recognize the successful tests performed in another jurisdiction. For example, even though Nader Khalili, in association with the ICBO and the Hesperia, California, building department, successfully passed static and dynamic load testing on two domes (one brick and one earthbag) in the highest earthquake zone in the country, these tests are not transferable to other jurisdictions. In other words, the same tests have to be performed again and again in each of the thousands of jurisdictions that exist in the US alone. The only recourse for changing these laws is to challenge them in court, a costly and time consuming process that the majority of owners/builders cannot afford or are loathe to pursue.

In spite of the restrictive nature of the codes in regards to alternative architecture, the determined builder can find ways to circumvent these apparent obstacles. Hundreds (if not thousands) of houses built without code compliance exist in the US without compromising the health and safety of their inhabitants. Many people choose to ignore the codes, noting that it is easier to beg forgiveness than to ask permission. The building department does not actively seek out code violators. Almost all code violations are brought to the attention of the building department by individual informants. Disaffected, hostile neighbors are the greatest source of these complaints. The obvious solution to this problem is to either live somewhere with no neighbors, or to approach them with neighborly intentions, which allows the opportunity to dispel any doubts or misconceptions. The bond that develops between people through open, honest discourse often makes converts of skeptics. Education is still our greatest tool in fostering change.

When building earthbag domes, we encourage people to begin by doing a small diameter dome. This addresses two separate matters. It allows someone new to dome building an opportunity to practice the techniques on a smaller project before attempting a larger one. It also enables you to build a structure that is small enough to not be affected by restrictive codes. Most building department jurisdictions do not require a permit if the "footprint" (or floor area) is less than what the code requires a permit for. In our case, a permit is not required if constructing a building with less than 120-square-feet (11.6 sq. m) of floor area. Check the local codes for the specific requirements in your area.

This is just one way of eluding the building codes. There are probably as many forms of evasion as there are people who make use of them. Our intention here is not to tell the public how to dodge their local building codes. While sidestepping the codes may appear to be the simplest way to build what we want, avoidance does absolutely nothing towards changing the code. If anything, continued clandestine evasion may hurt us in the long run by creating more limiting codes, stricter enforcement, and harsher penalties for the growing number of code violators. (Then again, the easiest way to bring speeding violators under control was to raise the speed limit.)

Changing the building codes is the only lasting recourse available to us. As deforestation continues to be a problem facing our society, the use of wood products for building will inevitably become more costly. A greater number of people are calling for more ecologically appropriate building materials, like straw bale and earth. With this in mind, let's look at some ways the code could address and accommodate the use of alternate materials.

The ICBO needs to add a new section to the existing codes. Standardizing the approval and inspection of alternative materials, like earth, would make getting a permit for an earthen home as easy as getting a permit for a frame house. Standardization is not as difficult as it may appear. Several building techniques, labor practices, and the materials themselves, have already been researched and time-tested. The ICBO need only acknowledge these facts. Banks and corporations have funded testing of new materials in the past (and present), but generally they expect some return from the sale of these new products. It is unlikely they would fund the testing of a natural material like earth when they foresee no way of marketing it. If funding for necessary tests were made available by government-funded programs, the need for making a profit would be superceded.

Tests for earth construction already exist, such as the example noted previously in this chapter. If the data from these tests were researched and gathered, they could be included in a new code section, thereby making redundant testing jurisdiction by jurisdiction unnecessary. Furthermore, why not accept the test of time demonstrated by earthen structures worldwide?

These buildings could be tested *in situ* to determine their strength, fire resistance, and safety. Currently, codes only recognizes tests done on individual building units, not the complete structure. In the case of earthbags, the monolithic nature of a dome is substantially stronger than its individual units. "United we stand, divided we fall" was a motto of the American Revolutionaries and an apt slogan for monolithic architecture.

Another way the building codes can relax their stranglehold on the public would be to allow the owner/builder to shoulder the legal responsibility for building with alternative materials. A "buyer beware clause" could be included in any title of an alternatively built structure offered for sale. Most owners/builders take the matter of safety very seriously for their own families, and probably would be willing to assume liability for their efforts. If building department officials would adopt the role of code facilitator rather than enforcer, more builders would view them as an enlightened friend instead of as an autocratic foe. The administrative branch of local government appoints building officials. What a difference it would make if the building official were an elected position. A well-informed, social-minded inspector would truly serve the public in regards to health and safety. No longer would you hear a building official's supercilious comment on alternative architecture, "It's not in the book."

As builders, we have the right and the responsibility to demand a change to outdated and environmentally irresponsible dictums. Whether it is through evasion, coercion, legislative means, or political manipulation by individuals or an organized lobby of homebuilders, buyers, and environmental activists, the codes will change. They cannot be supported by consumptive, unsustainable industry for much longer. We will (at worst) eventually reach the limit of our finite sources of timber, steel, and concrete. With any luck, the changes will occur before the total depletion of the natural resources needed to create these materials. Perhaps then the intent of the codes will reflect the need to protect the health and safety of the natural world that ultimately protects the health and safety of all people.

SANDBAG / SUPERADOBE / SUPERBLOCK: A CODE OFFICIAL PERSPECTIVE

(Condensed from Building Standards Magazine, *September-October 1998)* (Reprinted by permission).

When architect Nader Khalili first proposed constructing buildings made of earth-filled sandbags, the building department was skeptical. If we hadn't been trained to be courteous, we would have laughed out loud. How could anyone believe you could take native desert soil, stuff it into plastic bags and pile them up 15 feet (4.5 m) or more high? If they didn't fall down from their own weight, the first minor earthquake would cause a total collapse. How could a responsible building official condone such building code heresy?

Well, Nader Khalili is a very persistent man. Over time, he convinced us he was going to prove our skepticism wrong, that earth-filled sandbags (now called Superadobe) could meet the standards of the 1991 *Uniform Building Code™* (UBC). It started with Sections 105 and 107, allowing building officials to consider the use of any material or method of construction "... provided any alternate has been approved by the building official" and to require testing to recognized test standards as determined by the building official.

We contacted the International Conference of Building Officials (ICBO) Plan Review Services to see if they would perform the plan review for our city. ICBO welcomed the challenge, but indicated the same skepticism, since Hesperia, California, is within Seismic Zone 4 and local examples of this type of construction are nonexistent.

Negotiations resulted in a static load test program designed to add 200 percent of the UBC loading of 20 pounds per square foot (psf) (97 kg/m2) live and 20 psf (97 kg/m2) wind load. The first test used an 80 psf (390 kg/m2) loading of additional sandbags over one third of the exterior surface and, after monitoring, over one half of the exterior surface. During the entire test period, deflection was monitored to verify if ultimate loading was approached. Two domes, one of sandbags and one of unreinforced brick, were tested. Test results showed "that there was no movement of any surface of either dome structure as a result of the loading described in the test procedure." The domes had passed their first test.

After reviewing the test results, ICBO's Plan Review Services staff felt that the use of the domes should be limited to 15-foot (4.5 m) domes of Group M, Division 1 or Group B, Division 2 occupancies until sufficient monitoring had been completed. Mr. Khalili was principally interested in Group R occupancies, although he was also proposing the construction of a museum and nature center, a building that would house a Group A occupancy in a 50-foot (15 m) diameter dome. Mr. Khalili notified the city that he would not accept the size and occupancy limitations and would propose new testing to approve the use of larger structures.

After extensive negotiations, we agreed to a dynamic test procedure that involved applied and relaxed loads over a short period of time, with a series of tests with increasing loads until Seismic Zone 4 limits were exceeded. Tests involved three buildings, including the brick dome, the sandbag dome, and a sandbag vault structure with 5-foot-high (1.5 m) vertical walls and a barrel vault above. The tests were conducted and monitored by an ICBO-recognized testing laboratory in December 1995, and the required test limits were greatly exceeded. Testing continued beyond agreed limits until testing apparatus began to fail. No deflection or failure was noted on any of the tested buildings.

The plans went back to ICBO, and after final plan check comments were satisfied, ICBO recommended the plans for approval in February 1996. Our skepticism had long since vanished, as we had seen this style of building meet and exceed the testing of rational analysis as required by our code. Mr. Khalili had succeeded in gaining acceptance by the City of Hesperia for a building made of sandbags filled with earth. It is a testament to Mr. Khalili's perseverance and to the flexibility of the UBC.

— Tom Harp, Building Officer/Planning Director, City of Hesperia, California

— John Regner, Senior Plans Examiner, City of Hesperia, California

Build Your Own Dirtbag Tools

Bag Stands

The **Workhorse** is a rigid welded metal stand built specifically for each particular size bag (Fig. A.1). A simple formula for determining the dimensions for building a rigid metal stand for any size bag goes like this:

> Top ring of stand = 1 inch (2.5 cm) smaller than circumference of bag.
> Bottom ring of stand = 1.5 times the circumference of bag.
> Height of stand = 6 inches (15 cm) shorter than empty bag length for a 50-lb. bag and 9 inches (22.5 cm) shorter than empty bag length for a 100-lb. bag.
> **To find circumference of bag,** lay an empty bag flat and measure the width twice.

Our Favorite: The Collapsible Bag Stand

This is a "weld free" metal stand. It packs great in a suitcase! (Fig. A.2).

These directions are for the typical 50-lb. bag approximately 17 inches (42.5 cm) wide by 30 inches (75 cm) long. Cut two lengths of ½-inch (1.25 cm) or

¾-inch (1.875 cm) wide, 1/8-inch (0.3 cm) thick flat metal stock, 80 inches (200 cm) long. Bend at the designated dimensions beginning at the bottom. Fit one hoop snugly inside the other. Overlap the top seams and tape with duct tape. Drill the holes for the pivot-point one inch (2.5 cm) above center. Tighten snugly to give it enough friction to stay upright without collapsing.

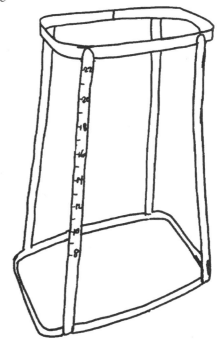

A.1

How to Make Your Own Collapsible Bag Stand

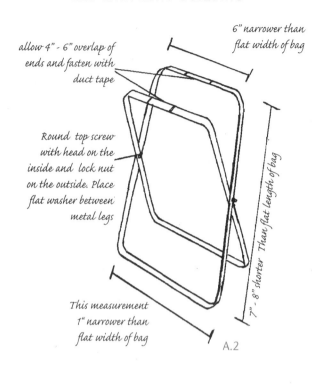

allow 4" - 6" overlap of ends and fasten with duct tape

6" narrower than flat width of bag

Round top screw with head on the inside and lock nut on the outside. Place flat washer between metal legs

7" - 8" shorter than flat length of bag

This measurement 1" narrower than flat width of bag

A.2

Cut ½-inch (1.25 cm) or ¾-inch (1.875 cm) wide by 1/8-inch (0.3 cm) thick flat metal stock to the appropriate length for the size bags you will be using. (When measuring the lay-flat width of a gusseted bag, be sure to open the folds to get the actual width of the bag.) For example: For a 17-inch (42.5 cm) wide by 30-inch (75 cm) long 50-lb. bag (laid flat), cut two pieces of flat metal stock about 80 inches (200 cm) long each. Mark the center point of the metal (Fig. A.3).

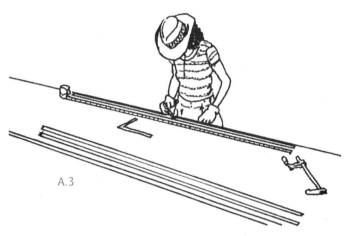

A.3

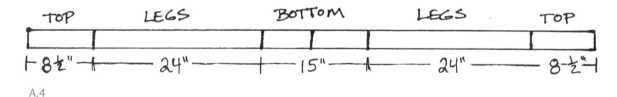

TOP   LEGS   BOTTOM   LEGS   TOP

8½"   24"   15"   24"   8½"

A.4

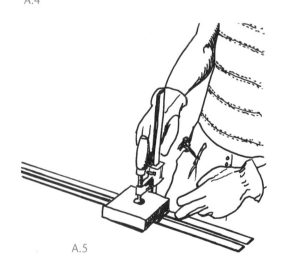

A.5

Measure the bottom width of the bag stand from the center point and mark it. Then measure the length of the legs from the marks indicating the bottom width of the bag stand. Mark these two points. The remaining eight inches (20 cm) at the ends will become the top of the bag stand including about three inches (7.5 cm) on each end for overlap (Fig. A.4).

Clamp the metal stock at the inside of the marks closest to the ends first (Fig. A.5)

Bend each end up at a strong 90° angle (Fig. A.6).

Now place the clamps inside the designated marks on the metal stock and bend them up at a strong 90° angle (Fig. A.7).

Do this on both sides (Fig. A.8)

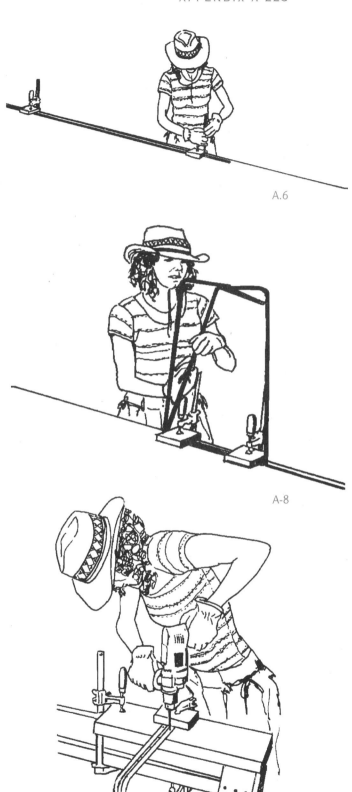

A.6

A.7

A-8

To find the pivot point, measure up from the bottom, half the height and add two to three inches (5 cm-7.5 cm). Mark it. This is done so that when the stand is operating, the bottom is wider than the top (Fig. A.9).

Drill a hole at this mark with a bit sized for a #10, ¾-inch (1.875 cm) long, slotted head machine screw for the ½-inch (1.25 cm) wide metal stock and/or a larger hex head bolt for the ¾-inch (1.875 cm) wide metal stock (Fig. A.10).

A-9

A-10

A-11

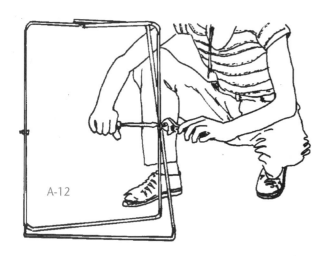

A-12

A-14

Paint the metal to help keep it rust free (Fig. A.11).

Place one metal frame inside the other. Connect the two frames at the pivot point by inserting the machine screw with the slotted head to the inside of the frame. Place a washer in between the two legs. Fasten an appropriate-size stop nut (a nut with nylon bushing inside of it) onto the end of the machine screw. Tighten with a screwdriver and a wrench (Fig. A.12).

To adjust the top width of the stand, overlap the ends until the width is about six to eight inches (15-20 cm) narrower than the lay flat width of the bag. The narrower the top of the bag stand is adjusted, the wider it will be able to spread open. Tape the overlap together with duct tape to a width that will create an opening that a #10 coffee-can will easily fill (a seven to eight inch [17.5-20 cm] spread is usually ideal) (Fig. A.13).

For added convenience during construction, mark each side of one of the legs on the bag stand with one inch (2.5 cm) measurements. Holding a square upright next to the open bag stand will help make the measurements accurate (Fig. A.14).

A-13

Homemade Concrete Tampers

Full Pounders

A six-inch (15 cm) diameter plastic planter pot filled six inches (15 cm) deep with concrete will make a pounder that weighs about 13 pounds (6 kg). This makes a pretty comfortable weight tamper. Cut the bottom out of the pot and turn it upside down.

A hardwood dowel or wild harvested hardwood about 1.25 inches (3.125 cm) thick and 4 feet (2.4 m) long makes a strong comfortable handle (Fig. A.15).

Starting half an inch (1.25 cm) from the bottom of the handle, pre-drill three ¼-inch (0.625 cm) holes at alternate angles about 1.5 inches (3.75 cm) apart. Tap in three sections of ¼-inch (0.625 cm) thick steel rod (all thread works fine). Be sure to cut the steel rods short enough to fit inside the tapered shape of the planter pot form without touching the sides of it. Twist a mess of barbed wire around the rods to give tensile strength to the concrete. Slip the pot over the top of this prepared handle (Fig. A.16).

A-15

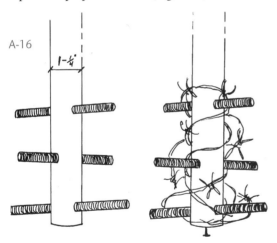

A-16

Quarter Pounders

Quarter pounders are built in the same manner as the full pounders, with three rods of steel and barbed wire for tensile strength. Use the one quart (0.94 liters) size yogurt containers as the forms, or any plastic quart size container that has a nice tapered shape. Because the quarter pounder form is used right side up, you can leave the bottom intact. Handle lengths vary according to the use of the pounder. The two-footer

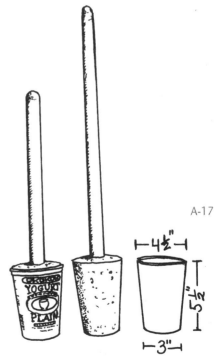

A-17

(1.2 m) for tamping keystone bags, and a standing height for tamping hard-ass bags works well (Fig. A.17).

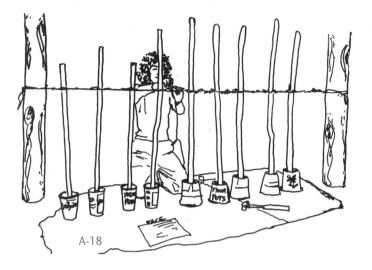

A-18

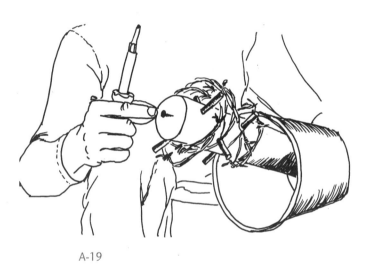

A-19

A-20

Suspend the handle from a taut rope (Fig. A.18). The handles should be elevated about one half inch (1.25 cm) above the ground, or drive in a screw to raise the handle off the ground about half an inch (1.25 cm). This will allow the concrete to fill in the gap below the bottom of the handle (Fig. A.19).

Prepare a rich mix of one part cement to two parts coarse washed concrete sand, stiff but not dry. Ladle the concrete into the form making sure to jiggle out any air bubbles by poking the mix inside the form with a stick. Tapping the exterior of the form also helps release trapped air.

Fill the form up to six inches (15 cm). Set a weight on top of the form to prevent the concrete from oozing out the bottom. A stiff mix will stay in there pretty good. Let it cure for one day. Remove the screw from the bottom. Slow cure the tamper with repeated wetting for several days. A week's worth of cure ensures a strong bond (Fig. A.20).

Window and Door Forms

This is the area of earthbag building that uses the most wood. The good thing about it is that the forms can be reused, sold, leased, loaned, or reconstructed into furniture or shelving. Forms need to be built deeper than the width of the bag wall to prevent the bags from wrapping around the edges during construction. Construct box forms large enough to accommodate the rough opening sizes needed for installation of windows and doors.

Plywood Sheathed Box Forms

Using two-by-fours for corner bracing works well, but any dimensional lumber will do. Install a solid plywood sheet on the bottom and four sides of the form. 5/8-inch (1.5 cm) or ¾-inch (1.875 cm) plywood contributes to making a sturdy form with ample shear strength. The top of the form can be open-spaced boards. Be sure to install adequate blocking inside the boxes to prevent distortion from ramming the bags up against them during construction. Cut out handholds for removing the forms (Fig. A.21).

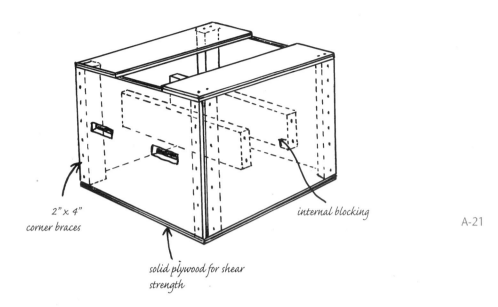

2" x 4"
corner braces

internal blocking

A-21

solid plywood for shear
strength

A *split-box form* is built in two halves with wedges placed in between them for easy removal from a finished wall.

Constructing them is still the same as the above-described process (Fig. A.22). Build them in even foot measurements for simplicity and reusability. Various dimension wedges can be used in between to accommodate odd width window and door sizes.

A full size, *mineshaft style* door form can be constructed using two sets of four-by-four posts framed with two-by-fours at top and bottom, and sheathed on either side with plywood. This type of form can be dismantled after the wall reaches door height and reconstructed into an overhead lintel (Fig. A.23).

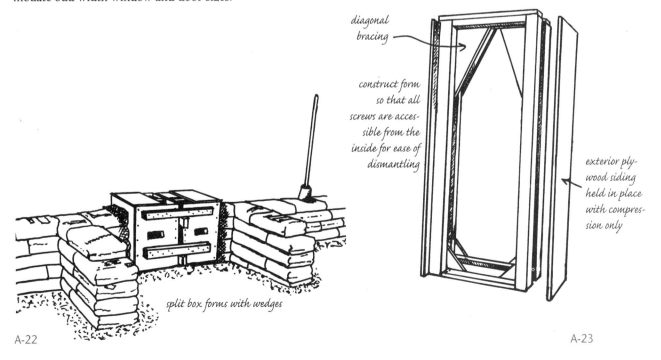

diagonal
bracing

construct form
so that all
screws are acces-
sible from the
inside for ease of
dismantling

exterior ply-
wood siding
held in place
with compres-
sion only

split box forms with wedges

A-22

A-23

Plywood-End Roman Arch Forms

Scribe a circle on a sheet of 5/8-inch (1.5 cm) or ¾-inch (1.875 cm) thick plywood.

Split it down the center (Fig. A.24 & A.25).

This provides both faces of the arch. Cut enough two-by-four ribs to be evenly spaced at most nine inches (22.5 cm) on center over the entire arch form. Top and bottom ribs should be placed narrow edge out. The rest should be installed the opposite way. The bottom can be solid plywood or boards.

Skin the top of the arch with 1/8-inch (0.3 cm) Masonite or any sturdy yet bendable substitute. Begin screwing from the top of the form and work toward the bottom ends one rib at a time. Sink screws deep enough to prevent them from catching onto the

A-24

A-25

chicken wire cradles when removing the form from the wall. Cut handholds into the two faces of the arch form for ease of removal and carrying (Fig. A.26).

Plywood-End Gothic Arch Forms

Gothic arches are designed in a variety of styles. They all share the same common shape of being more steeply sided than Roman arches. Creating a template on cardboard can aid in making the plywood ends, especially if multiple forms of the same shape are desired. See Chapter 10 for directions on creating a Gothic (or Egyptian) shaped arch. Follow the same procedure as a Roman arch for making the ribs. The skin for this type of arch can be made from two pieces of Masonite or a sturdy, bendable substitute (Fig. A.27).

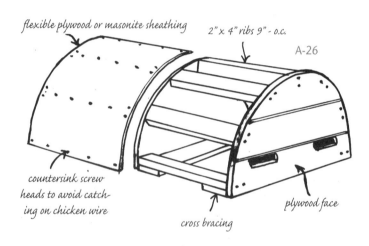

flexible plywood or masonite sheathing

2" x 4" ribs 9" - o.c.

A-26

countersink screw heads to avoid catching on chicken wire

cross bracing

plywood face

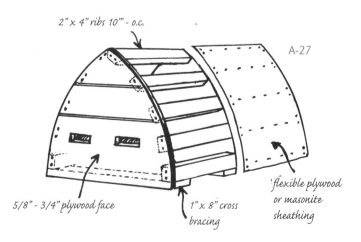

2" x 4" ribs 10" - o.c.

A-27

5/8" - 3/4" plywood face

1" x 8" cross bracing

flexible plywood or masonite sheathing

Solid Wood Forms (No Plywood)

Box and arch forms constructed from two-by lumber and boards are more laborious to build, but when built from discarded pallets, they can be constructed for almost nothing (Fig. A.28).

Box forms made this way should follow the same criteria as for plywood forms. They should be strong enough to resist the forces applied from compacting bags against them. They should also be diagonally braced to resist shear forces.

Dismantle pallets by sawing the nailed boards with a Sawsall, or pry them apart using a big wrecking (crow) bar. Making a template of the intended arch shape out of sturdy cardboard allows one to cut the desired shape of the form. The following are some examples of forms built with one-by and two-by dimensional lumber (Figs. A.29, A.30, A.31 & A.32).

A.28: *Discarded pallets.*

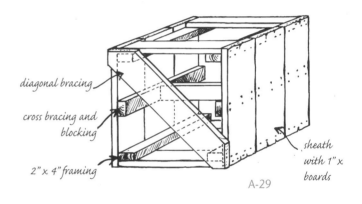

diagonal bracing

cross bracing and blocking

2" x 4" framing

sheath with 1" x boards

A-29

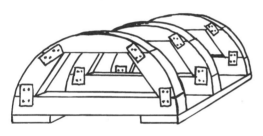

A-30

use 3 trusses for long forms or just two for forms under 2' long

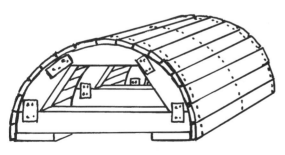

sheath with boards

A-31

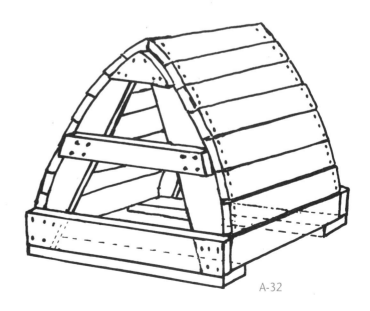

A-32

Wedge Box for Making Fan Bags

The two front slots on the face of the wedge box are cut at an angle that produces a fan bag shaped to fit any size Roman arch. The overall height of the box is 12 inches (30 cm). Make the box three inches (7.5 cm) longer than the width of the largest bags you will be using (Fig. A.33). Please read how to use a wedge box in Chapter 3.

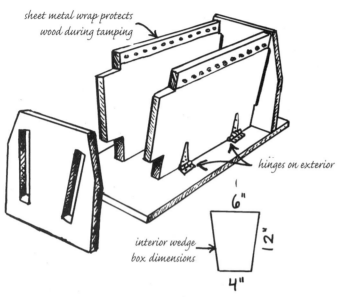

sheet metal wrap protects
wood during tamping

hinges on exterior

interior wedge
box dimensions

6"

12"

4"

A.33: The Wedge Box

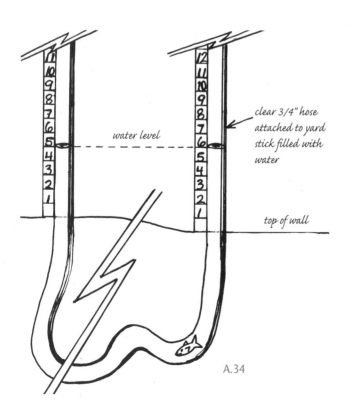

water level

clear 3/4" hose
attached to yard
stick filled with
water

top of wall

A.34

Make Your Own Water Level

To make a water level requires two yardsticks that can be obtained at most hardware stores, and a long piece of clear plastic tubing. The tubing should be long enough to span the greatest length of your structure. Attach either end of the clear tube to the two yardsticks. Attach them about two feet (60 cm) up the yardstick. Place the two yardsticks with the attached tube next to each other on a level surface and carefully fill the tube with water until the level of the water is readable along the yardsticks. Water does seek its own level, and if the yardsticks are on the same surface next to each other, they will register at the same height. If they don't, air is probably trapped somewhere in the tube and the air bubble needs to be chased out to one end.

Once the water is level at both ends, measurements can be taken at two different locations to check their level. Designate one location as your point of reference. When checking the level of different locations on the same wall, it's always in comparison to your point of reference. The trick to remember here is that the lower surface will read higher on the water level than the higher surface. The water will be level with itself so, for example, if the water level shows a one-inch (2.5 cm) difference between the two sides, the side with the higher reading is actually lower (Fig. A.34).

By raising the yardstick that reads highest by one-inch (2.5 cm), the water level on that side drops one-half inch (1.25 cm), and on the other end, the water level rises one-half inch (1.25 cm). Try this for yourself. This is definitely a case where seeing is believing, and is much more understandable to do than to read.

Sliders

Sliders are used after the first row of barbed wire is laid so the bags can be maneuvered into position until ready to be "Velcroed" into place. Any rigid sheet metal will do. These can often be obtained for free from ventilation fabrication shops. This size is a general all-purpose size for using with a 50-lb. bag (Fig. A.35).

Round off the ends and file the perimeter for safe handling. Using pliers, bend one edge of the slider twice to provide a smooth finger grip for pulling them out from under the bags during construction. Safety first. Make a variety of sizes for a variety of uses.

Barbed Wire Holders

Here are some examples of barbed wire holders, but by no means are they the only examples. Use your imagination and the materials on hand to create your own (Fig. A.36 & A.37).

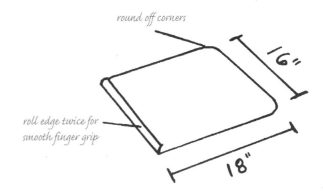

round off corners

roll edge twice for smooth finger grip

16"

18"

A.35: *Sheet Metal Slider*

A.37: *Buck stand barbed wire holder*

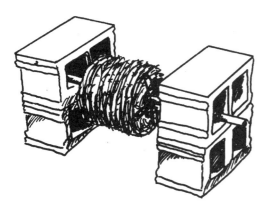

A.36: *Cinder block barbed wire holder*

How To Figure Basic Earthbag Construction Costs, Labor, and Time

Let's say we want to build a wall 9 feet (2.7 m) high and 100 feet (30 m) long. (9' (2.7 m) x 100' (30 m) = 900 square ft. [81 sq. m] of wall).

In the dirt bag world, we call a typical woven polypropylene feed bag that holds 50 pounds of grain, a *50-lb. bag.* When empty and laid flat, the average 50-lb. bag measures 17 inches (42.5 cm) wide by 30 inches (75 cm) long. When filled and tamped with dirt we call it a *working 50-lb. bag* and it measures 15 inches (37.5 cm) wide, 20 inches (50 cm) long and 5 inches (12.5 cm) thick. Now that we know the size of our bag ...

How many bags will we need for this wall?

Convert bag height and length into feet. Divide height of wall by .42 ft. (12.8 cm) (thickness of bag) and length of wall by 1.67 ft. (50.9 cm) (length of bag).

Height: 9 ft. (2.7 m) ÷ .42 ft.(12.8 cm)/bag = 22 rows of bags.

Length: 100 ft. (30 m) ÷ 1.67 ft. (50.9 cm)/bag = 60 bags per row.

Now multiply 22 x 60 = 1,320 bags for total wall.

How to figure costs of materials per bag:

Costs vary according to your particular situation. Since individual circumstances determine the cost per bag, let's do a case study on our particular scenario:

The bags usually come in bales of 1,000. We paid $140.00 per bale: $.10/bag and $40.00 for shipping.

The dirt.

We purchased "reject sand" at $1.25 per ton in a 15-ton truckload with a $35.00 delivery charge. $1.25 per ton x 15 tons = $18.75 + $35.00 (for trucking) = $53.75 total for 15 tons of dirt. Now divide $53.75 by 15 tons = $3.58 per ton.

So: How many 50-lb. bags per ton?

A single working 50-lb. bag weighs approx. 100 pounds.

There are 2,000 pounds in a ton. 100 pounds goes into 2,000 pounds 20 times. That makes 20 bags per ton. It figures from this that a 15-ton truckload will fill approximately 300 bags. Therefore, the amount of dirt needed to fill 1,320 bags for this particular project would be 66 tons.

The wire:

Our average cost for a 1,320-ft roll of heavy gauge, 4-point barbed wire is $50.00. To figure the amount of wire, we know there are 22 rows of bags, each row 100 feet (30 m) long. 22 x 100 = 2,200. Add 15% to this figure to account for overlap and wastage. 2,200 + (0.15 x 2,200) = 2,530 ft. of wire. 2,530/1,320 = approx. 2 rolls of wire.

Add up all the goods:

1,320 bags at $0.14/bag = $184.80
66 tons of dirt at $3.58/ton = $236.28
2 rolls 4-point barbed wire at $50/roll = $100.00

Total = $521.08

(Add 15% to this total figure for waste and miscellaneous items)

$521.08 + 15% = $599.24 or approximately $600.00.

Calculate costs of materials per square foot like a real contractor:

Now take our $600 and divide it with our 900 sq. ft. (86.4 sq. m) of wall and we get close to $0.67 per square foot ($6.94 per square meter) for the basic dirtbag system.

Labor costs:

Sure, the materials are cheap but the labor must be astronomical! Let's see…
Figure out how much wall gets built in one hour per person. Think of the wall as a whole system, not just a bunch of bags flopped on top of each other. We approximate a *conservative over-all time of four bags per hour per person for the entire construction of this wall.*

This figure includes: filling wheelbarrows and bags, laying wire, leveling forms, hard-assing, tamping, installing cradles and strip anchors, and slowing down as walls get higher and cans tossed farther.

Take total number of bags and divide by number of finished bags per hour:

This gives us our people-hours:
Example:
1,320 bags divided by 4 per hour = 330 hours.

Let's say our wage is $12.00 per hour. $12.00 x 330 = $3,960 in labor to build 900 sq. ft. (87 sq. m) of wall.

Cost of materials plus labor

Now let's add our materials: $600.00
Labor at $12.00 per hour: $3,960.00
Total $4,560.00

Square footage for materials and labor:

$4,560.00 divided by 900 square feet (87 sq. m) = $5.07 per square foot ($52.41 per square meter) for materials and labor. *Above costs reflect a site-specific-scenario.*

How long will it take?

People hours for this kind of construction are most effective with several teams of two, with a couple of rotating crew members to keep the wheelbarrows full. Six people in three teams of two with a seventh person delivering dirt could lay about 24 bags an hour, or 192 bags in an eight -hour day, and the above-mentioned 1,320 bags in 7 days. Adding additional crew members as the wall gets taller or for building a dome keeps the pace moving quickly and efficiently. Getting the walls up with help, especially as you get higher, keeps one's spirits high as well. The quick progress of such an oh-so-strong wall is exciting.

Other Cost Considerations

Foundations: Rubble trench, rammed earth tires, conventional concrete, etc.

Temporary wooden box and arch forms: These are reusable components that can be used for multiple projects, rented, resold, or used in trades.

Finish plasters: Vary from local adobe earth/straw/clay with optional lime plaster, to conventional cement stucco anchored to chicken wire.

The design of the structure: The bigger and more complex the design, the higher the cost.

Keep projects practical, compact, and FQSS!

Conversions and Calculations

Definitions

Diameter (D) = the width of a circle
Radius (r) = half the width of the diameter
Circumference (C) = the perimeter length of a circle
Area (A) = the square footage of a circle
Pi (π) = (3.1416)

To Calculate the Area and Circumference of a Circle

Example:
To find the area of a 20-foot (6 m) diameter circle:
π (3.1416) x Radius x Radius or: $A = \pi \ (r \times r)$
(Multiply *pi* (3.1416) x radius x radius (10ft. [3m] x 10ft. [3 m]) = 314.16 square feet (94.38 sq. m) of floor space.

Example:
To find the circumference of a 20-foot (6 m) diameter circle:
Circumference $= \pi (3.1416)$ x diameter or: $C = \pi \times D$
Multiply *pi* (3.1416) x Diameter (20 ft. [6 m]) = 62.8 (18.8 m) perimeter feet

Inch to Foot Conversion Table

1" = 08'
2" = .17'
3" = .25'
4" = .33'
5" = .42'
6" = .50'
7" = .58'
8" = .67'
9" = .75'
10" = .83'
11" = .92'
12" = 1.00'

Metric Conversions

1 inch = 2.5 centimeters
12 inches = 30 centimeters
1 foot = .3 meter
1 gallon = 3.75 liters
2.2 lbs = 1 kilogram
10.34 sq. feet = 1 sq. meter

One working 50-lb. bag is closely equivalent to 0.7014 square feet (0.065 sq. m) of wall surface (5 inches [12.5 cm] thick by 20 inches [50 cm] long).

One working 100-lb. bag is closely equivalent to 1 square foot (0.09 sq. m) of wall surface (6 inches thick [15 cm] by 24 inches [60 cm] long).

The Magic of a Circle

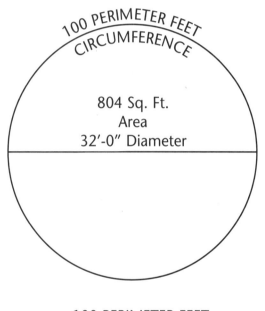

100 PERIMETER FEET
CIRCUMFERENCE

804 Sq. Ft.
Area
32'-0" Diameter

Nature is more than a structural engineer; She is also an expert in energy efficiency. A round wall uses the least amount of materials while providing the maximum amount of space. By trading corners for curves we fortify the structural integrity of our architecture while rediscovering our intuitive understanding of nature's dynamic engineering principles.

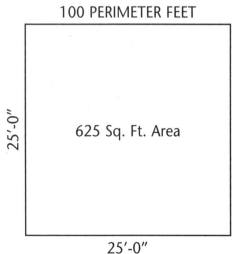

100 PERIMETER FEET

25'-0"

625 Sq. Ft. Area

25'-0"

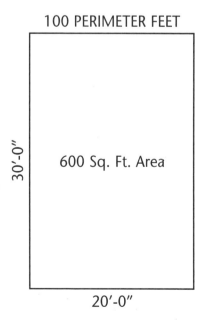

100 PERIMETER FEET

30'-0"

600 Sq. Ft. Area

20'-0"

Resource Guide

Books

Allen, Edward. *Stone Shelters*. MIT Press, 1981.

Bee, Becky. *The Cob Builder's Handbook*. Groundworks, 1997.

Boily, Lise and Jean-Francis Blanchette. *The Bread Ovens of Quebec*. National Museums of Canada, 1979.

Bourgeois, Jean-Louis and Carol Lee Pelos. *Spectacular Vernacular*. Aperture Foundation, 1996.

Chiras, Daniel D. *The Natural House*. Chelsea Green Publishing, 2000.

Courtney-Clarke, Margaret. *African Canvas*. Rizzoli, 1990. (West African women's stunning vernacular art and architecture)

Easton, David. *The Rammed Earth House*. Chelsea Green Publishing, 1996.

Elizabeth, Lynne and Cassandra Adams. *Alternative Construction*. John Wiley and Sons, 2000.

Evans, Ianto, Linda Smiley, and Michael Smith. *The Hand Sculpted House*. Chelsea Green Publishing, 2002.

Ferguson, William M. and Arthur H. Rohn. *Anasazi Ruins of the Southwest in Color*. University of New Mexico Press, 1994.

Gray, Virginia, Alan Macrae, and Wayne McCall. *Mud, Space, and Spirit*. Capra Press, 1976.

Guelberth, Cedar Rose and Dan Chiras. *The Natural Plaster Book*. New Society Publishers, 2003.

Higa, Teruo, author; Anja Kanal, translator. *An Earth Saving Revolution II*. Sunmark Publishing, 1998.

Holmes, Stafford and Michael Wingate. *Building With Lime*. Intermediate Technologies Publications, 1997.

Houben, Hugo and Hubert Guillaud. *Earth Construction*. Intermediate Technology Publication, 1994.

Kahn, Lloyd, ed. *Shelter*. Shelter Publications, 1973. (This is a classic that should be on everyone's bookshelf)

Kemble, Steve and Carol Escott. *How to Build Your Elegant Home with Strawbales Manual*. Sustainable Systems Support, 1995.

Kern, Ken, Ted Kogon, and Rob Thallon. *The Owner-Builder and the Code*. Owner-Builder Publications, 1976.

Khalili, Nader. *Ceramic Houses and Earth Architecture*. Burning Gate Press, 1990.

Khalili, Nader. *Racing Alone*. Harper and Row, 1983.

Lime Stabilization Construction Manual. Bulletin 326, the National Lime Association.

Ludwig, Art. *Create an Oasis with Greywater*. Oasis Design, 1997.

Magwood, Chris and Peter Mack. *Straw Bale Building*, New Society Publishers, 2000.

McHenry, Jr., Paul Graham. *Adobe and Rammed Earth Buildings*. University of Arizona Press, 1984.

McHenry, Jr., Paul Graham. *The Adobe Story: A Global Treasure*. The American Association for International Aging and the Center for Aging, 1996.

Minke, Gernot. *Earth Construction Handbook*. WIT Press, 2000.

Nabokov, Peter and Robert Easton. *Native American Architecture*. Oxford University Press, 1989.

Pearson, David. *The Natural House Book*. Simon and Schuster/Fireside, 1989.

Reynolds, Michael. *Earthship*. Solar Survival Press,1993.

Rudofsky, Bernard. *Architecture Without Architects*. University of New Mexico Press, 1990 (3rd printing).

Rudofsky, Bernard. *The Prodigious Builders*. Harcourt Brace Jovanovich, 1977.

Smith, Michael G. *The Cobber's Companion*. Cob Cottage, 1998.

Soltani, Atossa and Penelope Whitney, eds. *Cut Waste Not Trees*. Rainforest Action Network, 1995.

Steele, James. *An Architecture for People (The complete works of Hassan Fathy)*. Whitney Library of Design, 1997.

Steen, Athena Swentzell, Bill Steen, and David Bainbridge. *The Strawbale House*. Chelsea Green Publishing, 1994.

Taylor, John S. *A Shelter Sketchbook*. Chelsea Green Publishing, 1997.

The Underground Space Center, University of Minnesota. *Earth Sheltered Housing Design.*, Van Nostrand Reinhold, 1979.

Tibbets, Joe. *The Earthbuilder's Encyclopedia*. Southwest Solaradobe School, 1989.

Williams, Christopher. *Craftsmen of Necessity*. Vintage Books, 1974.

Wojciechowska, Paulina. *Building with Earth.* Chelsea Green Publishing, 2001.

Wright, David. *Natural Solar Architecture.* Litton Educational Publishing, 1978.

Periodicals

The Adobe Builder Inter-Americas. Joe Tibbets, ed. P.O. Box 153, Bosque, NM 87006, Tel. 505-861-1255, e-mail: adobebuilder@juno.com, www.adobebuilder.com. (Quarterly featuring earthen architecture throughout the Southwest; offers two books that highlight recent amendments to building codes pertaining to adobe and rammed earth through Southwest Solaradobe School.)

Adobe Journal. Michael Moquin, ed. P.O. Box 7725, Albuquerque, NM 87194, Tel/Fax: 505-243-7801. (Out of print quarterly periodical - worth trying to find back issues.)

Building Standards: Trade Magazine of the International Conference of Building Officials. 5360 Workman Mill Rd., Whittier, CA 90601, Tel: 562-699-0541, Fax: 562-699-8031. (Heady stuff, but occasionally covers alternative construction methods.)

Communities-Journal of Cooperative Living. Diana Leafe Christian, ed. 52 Willow St., Marion, NC 28752, 828-652-8517, e-mail: communities@ic.org, www.ic.org. (Published 5 times per year)

Earth Quarterly. Gordon Solberg, ed. P.O. Box 23, Radium Springs, NM 88054. (Another out of print quarterly; try contacting the Solberg's for back issues.)

Environmental Building News. Nadav Malin, ed. 122 Birge St., Suite 30, Brattleboro, VT 05301, Tel: 802-257-7300, e-mail: ebn@BuildingGreen.com, www.BuldingGreen.com. (Lots of information on products and materials being developed for "green" users.)

The Last Straw Journal. The Green Prairie Foundation for Sustainability, P.O. Box 22706, Lincoln, NE 68542, Tel: 402-483-5135, Fax: 402-483-5161, e-mail: thelaststraw@thelaststraw.org, www.thelaststraw.org. (Quarterly publication - excellent resource for strawbale and beyond.)

The New Settler Interview. Beth Bosk, ed. P.O. Box 702, Mendocino, CA 95460, Tel: 707-937-5703. (Not much information on alternative building, but lots of other alternative information.)

RMI Solutions. Cameron M. Burns, ed. Rocky Mountain Institute, 1739 Snowmass Creek Rd., Snowmass, CO 81654. (Excellent resource for cutting edge ideas for a sustainable society and world.)

Earthbag Building Supplies and Products

Bags and Tubes

There are hundreds of sources of bag manufacturers and these are just a few. Check the web and the Thomas Register at your local library. Shop around and compare prices.

Cady Industries, Inc., P.O. Box 2087, Memphis, TN 38101, Tel: 901-527-6569, 800-622-3695. www.cadyindustries.com. Offer a wide variety of different size misprint bags (including gusseted and burlap), tubes on a roll, and other specialty items.

PolyTex Fibers Corporation, 9341 Baythorne Dr., Houston, TX 77041, Tel: 713-690-9055, 800-628-0034. www.polytex.com. Offers misprint, regular poly and gusseted, burlap, and tubes on a roll.

Kansas City Bag Co., 12920 Metcalf Ave., Suite 80, Overland Park, KS 66213, Tel: 800-584-5666. Bag brokers and distributors.

Black Poly Irrigation Tubing

Check agricultural supply outfits or special order from local plumbing supplies or lumberyards. Also called "utility pipe." Use the flexible, thin-walled 80 psi. tubing, ¾ - 1 inch diameter, as an alternative weep screed for curved walls.

Bulk Casein

American Casein Company, 109 Elbo Ln., Burlington, NJ 08016, Tel: 609-387-3130, www.americancasein.com. Bulk casein, 50 lb. minimum.

National Casein Company, 601 W. 80th St., Chicago, IL 60620, Tel: 773-845-7300.

Bulk Citrus Thinner (D'Limonene)

Odor Control, Barbara Lang, P.O. Box 5740, Scottsdale, AZ 85261, Tel: 888-948-3956 5-gallon drum minimum orders.

Bulk Oxide Pigments

(Check local concrete and stucco supply outlets first)

Building for Health Materials Center, P.O. Box 113, Carbondale, CO 81623, Tel: 970-963-0437, 800-292-4838, www.buildingforhealth.com. One stop shopping for environmentally safe plastering and building materials.

Color and Abrasives, 248 W. 9210 S., Sandy, Utah 84070, Tel: 801-561-0870, 800-675-5930, www.colorandabrasives.com. Good prices on bulk oxide pigments.

Kremer Pigments, Inc., 228 Elizabeth St., New York, NY 10012, Tel: 212-219-2394, 800-995-5501, www.kremerpigments.com. Widest variety of imported pigments; also a source for kilogram-size casein powder.

Laguna Clay Company, 14400 Lomitas Ave., City of Industry, CA 91746, Tel: 626-330-0631, 800-452-4862, www.lagunaclay.com. Manufacturers of pottery clays and bulk earthen and mineral pigments.

The Natural Choice, BioShield Paint Co., 1365 Rufina Circle, Santa Fe, NM 87505, www.bioshieldpaint.com. Natural paints and a source for BioShield's Natural Resin Floor Finish.

Bulk Soaker Hose

Moisture Master, Aqua-pore Moisture Systems, 610 S. 80th Ave., Phoenix, AZ 85043, Tel: 602-936-8083, 800-635-8379, www.moisturemaster.com.

Peaceful Valley Farm Supply, , P.O. Box 2209, Grass Valley, CA, Tel: 888-784-1722, www.groworganic.com.

Bulk Sources for Clay-Rich Soil

Your building site; excavation of new construction sites; ponds and roads; road cuts; gravel yards ("reject" or "top-soil" piles); suppliers of "ball field clay" (excellent source for clean quality bulk earth plaster soil).

Enharradoras

CRG Designs, Cedar Rose Guelberth, Designs for Living, P.O. Box 113, Carbondale, CO 81623, Tel: 970-963-0437. Workshops on earthen plasters and natural paints and finishes.

Gourmet Adobe, Carole Crews, HC 78 Box 9811, Rancho de Taos, NM 87557,
   Tel: 505-758-7251, e-mail: seacrews@taosnet.com. Workshops on earthen plasters, alis, and decorative finishes.

Keely Meagan, P.O. Box 5888, Santa Fe, NM 87502, Tel: 505-421-3788,
   e-mail: keelymeagan@hotmail.com. Author of "Earth Plasters for Strawbale Homes." Offers workshops and consultation in earth plaster and cob.

Ok Ok Ok Productions, Kaki Hunter, 256 E.100 S., Moab, Utah 84532, Tel:435-259-8378, e-mail: okokok@frontiernet.net. One awesome babe who likes to get down, get dirty, get in the mud, and has a great time doing it, too!

Papercrete/Paperadobe Pioneers

Eric Patterson, 2115 Memory Ln., Silver City, NM 88061, Tel: 505-538-3625. Consultation.

Hartworks, Inc., Kelly and Rosana Hart, P.O. Box 632, Crestone, CO 81131,
   Tel: 719-256-4278, 800-869-7342, e-mail: office@hartworks.com, www.hartworks.com.

Mike McCain, P.O. Box 265, Columbus, NM 88029, Tel: 505-531-2201. Papercrete workshops; designs and builds papercrete mixing machines.

Philip Mirkin, PO Box 123, Dove Creek, CO 81234, Tel. 970-677-3600, email philipmirkin@hotmail.com, www.hybridadobe.com.

Sean Sands, P.O. Box 4, Grand Forks, BC V0H 1H0. Papercrete/paperadobe innovator.

Landscape Filter Fabric: for "French Drains"

Peaceful Valley Farm Supply, P.O. Box 2209, Grass Valley, CA, Tel: 888-784-1722, www.groworganic.com.

Plaster Sprayer

Mortar Sprayer, Nolan Scheid, P.O. Box 2952, Eugene, OR 97402, Tel: 541-683-4167,
   www.mortarsprayer.com. This plaster sprayer is imported from Mexico.

Pond Liner, Heavy Mil Plastic, and Waterproof Foundation Membranes

Check locally for sources for most of these items. Below is a specialty product.

Delta-MS, Cosella-Dorken Products, Inc., 4655 Delta Way, Beamsville, ON LOR 1B4, Tel: 905-563-3255, 888-4DELTA4, www.DeltaMS.com. Unique "Air-Gap/Drainage Membrane."

Real Goods, Gaiam, Inc., 13771 S. Highway 101, Hopland, CA 95449, Tel: 707-744-2100, 800-919-2400, www.solar.realgoods.com.

Pozzolans

Ground pumice, fired brick fines, etc, are available at most pottery suppliers and are referred to as "grog." Also check gravel yards in volcanic deposit rich areas.

Pumice

Copar Pumice Plant, P.O. Box 38, Espanola, NM 87532, Contact: Rick Bell, Tel: 505-929-0103, www.coparpumice.com. Bulk pumice direct from the mine.

Tensioner Devices, Poly Strapping, and Banding Tool Equipment

Carlson Systems. Call 800-325-8343 for nearest dealer. E-mail: webteam@systems.com.

Tie Wires: Commercially Made

Gemplers Catalog (industrial agricultural supplier), Tel: 800-382-8473, www.gemplers.com. Double looped "wire ties" and twisting tools; also a fancy barbed wire dispenser on wheels.

Yurt Roof Material

Advance Canvas, P.O. Box 1626, Montrose, CO 81402, Tel: 970-240-2111, 800-288-3190, www.AdvanceCanvas.com. Provides yurt roof components "à la carte."

Networking Resource Information

This is by no means a complete listing, but includes some of our favorites...

Black Range Lodge, Catherine Wanek and Pete Fust, Star Rt. 2, P.O. Box 119, Kingston, NM 88042, Tel: 505-895-5652, e-mail: blackrange@zianet.com, www.BlackRangeLodge.com, waterworks.strawbalecentral.com. Home of Black Range Films, hosts of Southwest Natural Building Colloquium.

CalEarth/Geltaftan Foundation, Nader Khalili, 10376 Shangri La Ave., Hesperia, CA 92345, Tel: 619-244-0614, e-mail: calearth@aol.com, www.calearth.org. Offers 1-day, multi-day, and apprentice training workshops on earthbag construction.

Canyon Springs Consulting, Alison L. Kennedy, 847 Wagner Ave., Moab, Utah 84532, Tel: 435-259-9447, e-mail: alisonlara@frontiernet.net. Consultation for business, non-profits, and owner-builders.

Cob Cottage Company, Linda Smiley and Ianto Evans, P.O. Box 123, Cottage Grove, OR 97424, Tel: 541-942-2005. Offers cob workshops nationwide.

HUD-Housing and Urban Development, go to www.huduser.org and search for the publication entitled, "Frost-Protected Shallow Foundations in Residential Construction," (April, 1993), or www.cs.arizona.edu/people/jcropper/desguide.html. This site has specifications for the Shallow, Frost-Protected Foundation Systems.

New Mexico Adobe and Rammed Earth Building Codes, State of New Mexico, Regulation and Licensing Dept., Construction Industries Division, CID, P.O. Box 2501, Santa Fe, NM 87504, Tel: 505-827-7030.

Ok Ok Ok Productions, Kaki Hunter and Doni Kiffmeyer, 256 East 100 South, Moab, Utah 84532, Tel: 435-259-8378 e-mail: okokok@frontiernet.net, www.okokok.org. Earthbag, earth plaster, lime plaster — have workshop? will travel! Entertaining presentations, and so much more.

Out of Nowhere, Easter Tearie, Darnaway, Forres IV36 OST Scotland/UK, Tel/Fax 44 1309 641 650, www.outofnowhere.com. Information on reciprocal roofs.

Sascha Gut on Werner Imbach Gut Ag, Industriestrasse, 24 CH-4313 Mohlin, Switzerland, Tel: 41 61 851 1646, e-mail: gut-ag@swissonline.ch. Reciprocal frame design built in Switzerland.

Sustainable Sources, Bill Christensen and Jeanine Sih Christensen, Austin, TX, www.greenbuilder.com. One-stop online resource center for sustainability: green building, sustainable agriculture, and responsible planning.

Sustainable Systems Support, Steve Kemble and Carol Escott, P.O. Box 318, Bisbee, AZ 85603, Tel: 520-432-4292, 520-743-3828, e-mail: primalpulse@yahoo.com. Strawbale, earthbag, permaculture, design, consultation, workshops — these folks do it all, and do it great!

With Gaia Design, Susie Harrington, P.O. Box 264, Moab, Utah 84532, Tel: 435-259-7073, e-mail: wgaia@earthlink.net, www.withgaia.org.

Index

scooching (soft-packing), 48

scoria, 59, 65, 209

sealers, 193, 195

shelving attachments, 100–102

silt, 14–15

sliders, 37, 84

soil preparation, and moisture content, 17–21

soil ratios. *See also* soils

    determining, 15–16

    optimal, 6–7, 13, 15, 16

        exceptions to, 16–17

soils. *See also* soil ratios

    choosing the best, 16

    components, 13–15

    imported, 16

    moisture content, 17–21

    and organic matter, 15

    stabilized earth, 57–59

    well-graded, 15

solar temperature control, 208

Southern Standard Building Code, 217

split-box form, 87

springline, 127, 137, 154

stabilized-earth mixes, 57–59

stains, 193

stair attachments, 102

stem walls

    concrete, 56–57

    insulated earthbag, 63–64

    stabilized earth, 57–59

    tire, 62–63

    traditional and alternative, 62–63

straw bales, 210–12

strip anchors, 30–31, 88–89, 100–101

structural design features (walls)

    barbed wire, 72

    height limitations, 70

    height to width ratio, 70–71

    interlock corners, 72

    lateral support, 70

    locking row, 73–74

    openings, 73

    post and beam, 74–76

    round walls, 71

    tube corners, 72

structural integrity, 9

stucco, 16, 26, 84, 185–86

subterranean structures. *See* bermed / buried structures

sunrooms, 208, 213

swales (drains), 66

T

tampers (full and quarter pounders), 38

tamping, 18–19, 82

teamwork, 45–47, 52, 89

techniques

    bag whacking, 39

    can tossing, 36–37

    closing in a row, 47

    hard-assing, 38–39, 45

    hard-packing, 48

    laying a coil, 44, 154–55

    scooching, 48

    using a pole compass, 48–51

temperature control, 9–10

tensile strength, 8, 24, 112, 141, 188

tension ring, 110, 118, 135, 157

tensioner device, 112

testing

    corbelling process, 146

    for moisture content, 20

    soils, 15–19

    for structural integrity, 9, 17

thatch. *See* living thatch

thermal flywheel effect, 9

thermal performance, 9–10

Tibbets, Joe, 57

tie wires, 26–27, 84

tile, 165, 200–201

tire stem wall, 62–63

tools

    architectural compasses, 48–51

    bag stand, 34

    brick weights, 39–40

    cans, 36

About the Authors

Kaki Hunter and Donald Kiffmeyer have been involved in the construction industry for the last 20 years, specializing in affordable, low-tech, low-impact building methods that are as natural as possible. After being introduced to sandbag construction by Nader Khalili in 1993, they developed the "Flexible-Form Rammed Earth Technique" of building with earthbags and have taught the subject and contributed their expertise to several books and journals on natural building.

Both authors grew up with a great fondness for nature. Kaki Hunter was raised in the film and theatre world, her father being a professional film actor and teacher. She worked as a professional actor in Europe and then Hollywood, becoming an award-winner along the way. Doni Kiffmeyer's love of entertainment developed into political street theatre in the late sixties and early seventies in response to the Viet Nam war. He then worked as a river guide, combining his love of nature with a natural propensity for entertaining. They met performing in a community theatre play and realized that their shared interests included building and sustainability. This prompted them to investigate alternative architecture until they built their first arch. — They were hooked! They both fell in love with the idea of being able to build arches and domes out of earth. From these innocent beginnings they were launched into the alternative building world where they were encouraged to share their experiences of dome building with more people.

Together, they have written a screenplay (1995) entitled, Honey's House. It concerns a single mother's plight of homelessness that is solved with the help of friends and — what else? — Earthbag building! Now they are extending their interest in earthen and alternative construction methods that inspire happy, healthy habitats in harmony with nature by sharing that which inspires them.

If you have enjoyed *Earthbag Building,* you might also enjoy other

BOOKS TO BUILD A NEW SOCIETY

Our books provide positive solutions for people who want to make a difference. We specialize in:

Sustainable Living • Ecological Design and Planning • Natural Building & Appropriate Technology

New Forestry • Environment and Justice • Conscientious Commerce • Progressive Leadership

Educational and Parenting Resources • Resistance and Community • Nonviolence

For a full list of NSP's titles, please call 1-800-567-6772 or check out our web site at:

www.newsociety.com

NEW SOCIETY PUBLISHERS